T0135502

Nils Blüthgen

Systems-biological approach
to Ras-mediated signal transduction

Logos Verlag Berlin

λογος

Bibliografische Information Der Deutschen Bibliothek

Die Deutsche Bibliothek verzeichnet diese Publikation in der Deutschen
Nationalbibliografie; detaillierte bibliografische Daten sind im Internet
über http://dnb.ddb.de abrufbar.

ISBN 3-8325-1346-9

Logos Verlag Berlin
Comeniushof, Gubener Str. 47,
10243 Berlin
Tel.: +49 (0)30 / 42 85 10 90
Fax: +49 (0)30 / 42 85 10 92
http://www.logos-verlag.de

Contents

1

General Introduction

The competence of cells to respond to signals from the external environment is a property considered to be a fundamental characteristic of life (Sauro and Kholodenko, 2004). Even relatively simple unicellular organisms make use of a variety of signals: bacteria detect nutrients and swim towards them, they can switch between different metabolic pathways depending on their nutrients available in the environment, react upon temperature changes, or sense light. Yeast cells find mating partners by secreting and sensing hormones. In multi-cellular organisms, the ability to send and receive a multitude of signals enables the single cell to influence the behavior of other cells. By this means multi-cellular organisms can orchestrate the development of tissues and organs, they can possess an immune system that can distinguish between own and alien cells, they can heal wounds, etc. The carrier of the information passed between cells are mainly hormones or small molecules that are either secreted and bind to receptors, or that can directly flow to neighboring cells via gap-junctions. Some signals are also communicated via surface molecules that directly stimulate receptors via cell-to-cell contact, which have been termed juxtacrine signaling.

The need for communication and to sense the surrounding gave rise to enormously complex intracellular signal transduction networks that sense signals, integrate them and control processes that allow cells to respond adequately. Molecules that are present at the membrane can bind to specific receptors that trigger signal transduction cascades composed of intracellular signaling molecules, which are often kinases that can modify their targets by phosphorylation. Mammalian cells possess more than 500 kinases (1.7% of the human genes, Manning et al. (2002)) of which a large fraction forms the mammalian signal transduction network. The human genome codes for about 150-200 phosphatases that dephosophorylate proteins (Alonso et al.,

2004), and approximately 30% of the human genes are subject to phosphorylation (Cohen, 2001). Although not all of these proteins are necessarily involved in signal processing, these numbers highlight the importance of signal processing and regulation due to post-translational modification in higher organisms.

In the last decades, a large body of information about the topology of signal transduction networks accumulated. However, it becomes clear that also quantitative information about the dynamics is very important to gain insight into the functioning of these networks. For instance, the ability of the bacterium *E. coli* to swim toward nutrients is not only provided by its potential to sense a nutrient and respond to it. It is additionally necessary that the network shows fast perfect adaptation, so that it only senses changes in and not absolute levels of nutrients (Alon et al., 1999). Also, in mammalian cells the duration of signals is often very important to determine the response (Marshall, 1995). Typically, the structure of the network does not fully determine the dynamics of it, but it may give rise to several distinct behaviors depending on the kinetic properties of the molecules (Blüthgen and Herzel, 2001).

The target of signal-transduction is often the transcriptional activation or inhibition of genes. Kinases at the end of signal-transduction cascades phosphorylate transcription factors and thereby translate the activities of kinases into activity of transcriptional regulators. However, it is often a puzzle by which transcription factors a specific signal transduction cascade controls a physiological response. This question is not easy to be answered, as the physiologically relevant response is mostly governed by the regulation of expression of a multitude of genes, and for most genes the transcription factors are not known that connect signal transduction to transcriptional activation or repression.

This thesis focuses on a mitogen-activated protein kinase (MAPK) cascade, the Raf-Mek-Erk signal transduction cascade, which is a chain of molecules downstream of Ras that transduces the presence of growth-factors at the cell membrane to the transcriptional regulation of about hundred genes. It this thesis the following questions are addressed:

- Is signal-transduction via this cascade switch-like? Is this property robust? What are the potential mechanisms for it?

- Can one unveil the biological processes regulated by this cascade from microarray data? Are the regulated processes different for physiological stimuli and for pathogenic stimulation with mutated Ras?

- Do the induced genes feed back into the signal transduction cascades?

- Is sequence information and the annotation of genes sufficient to predict the processes that are regulated by Ras-regulated transcription factors?

To address these questions systems-biological methods are used, which approach biological systems by analyzing the interaction between its parts. While traditional genetics and bioinformatics have brought insight into the functioning of a gene and the process a gene is involved in, systems biology tries to integrate the available data. It aims to take a networks perspective on the system by combining diverse sets of data from microarray studies, time-series data, physiological information using modeling and statistical investigations.

For about a decade, the MAPK signal transduction cascade downstream of the Ras molecule is subject of mathematical modeling (Huang and Ferrell, 1996). It has been shown for *Xenopus* oocytes that the dose-response curve of this cascade is sigmoidal, a property that has been termed ultrasensitivity. The next chapter reviews the possible mechanisms to generate ultrasensitivity; it analyzes the robustness of these mechanisms and discusses the impact of feedback loops on the dynamics of ultrasensitive cascades. In the subsequent chapters the question whether ultrasensitivity is an intrinsic property of the MAPK cascade also in mammalian cells is addressed from two different sites: first, kinetic details are analyzed to understand under which conditions ultrasensitivity can arise; secondly, single-cell measurements are used to investigate whether the MAPK-cascade is ultrasensitive or even bistable *in vivo*.

The second part of this thesis addresses the transcriptional regulation by Ras-mediated signal transduction. First, using time-resolved measurements and mathematical modeling it is investigated how transcriptionally regulated genes feed back into the signal transduction process. Then a more global perspective is taken and the relevant physiological responses regulated by Erk are investigated.

Next, the biological background of Ras-mediated signal transduction shall be introduced, followed by an overview of the current state of modeling Ras-mediated signal transduction and the current view on transcriptional regulation by Ras.

1.1 Ras-mediated signal transduction

Ras activation

Ras-mediated signal transduction connects diverse extracellular stimuli to the genetic program of mammalian cells (see Fig. 1.1). Ras molecules, small

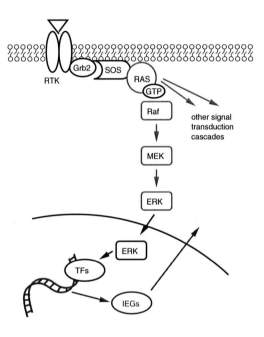

Figure 1.1: Sketch of signal transduction mediated by Ras . Hormones (triangle) bind to the receptor tyrosine kinase (RTK) which promotes the activation of Ras using adaptor proteins (in this case SOS and Grb2). Ras, in turn, anchors Raf to the membrane, where it is activated and phosphorylates Mek. Mek phosphorylates Erk, which translocates to the nucleus and activates transcription factors (TFs) that promote the expression of immediate-early genes (IEGs). The cascade composed of Raf-Mek-Erk is also called mitogen-activated protein kinase cascade (MAPK-cascade). It is a signal transduction cascade that is highly conserved among eukaryotes.

Table 1.1: Point mutations in molecules of the Ras-Raf-Mek-Erk cascade are involved in several types of cancers (Kufe et al., 2003).

Molecule	Cancers
H-Ras	Thyroid carcinomas, bladder carcinomas
K-Ras	Pancreatic, lung and colon adenocarcinomas, non-small-cell lung carcinoma, myeloid leukemias, thyroid carcinomas
N-Ras	Melanomas, myeloid leukemias, thyroid carcinomas
B-Raf	Melanomas, colon carcinomas, small-cell lung cancer

GTP-binding proteins, are membrane-bound intracellular signaling molecules that mediate a wide variety of cellular functions, including proliferation, differentiation, and survival. They act as a molecular switch alternating from an inactive GDP-bound state to an active GTP-bound state (Campbell et al., 1998). Depending on the cell type, the activation of Ras molecules can promote cell cycle progression, e.g. in fibroblasts, and differentiation in others cells, such as PC12 cells (Marshall, 1995). In healthy cells, the activation of Ras is mainly carried out by tyrosine receptors that are stimulated by growth factors. Upon stimulation with growth factors, the receptors recruit guanine nucleotide exchange factors (GEF) to the membrane and promote the release of GDP from the catalytic pocket of Ras. Subsequently, GTP that is available in access over GDP can enter this pocket and activate Ras. These processes result in a transient activation of Ras and its downstream signal transduction cascades. The timing and the relative activation of these downstream cascades is tightly controlled and determines the cellular response (Marshall, 1995; Asthagiri et al., 2000).

Ras-mediated signal transduction has been subject to intense investigations by many groups, as Ras and downstream molecules are major oncogenes (see Table 1.1). It is mutated in 15%-30% of human tumors, and in some cancers such as pancreatic carcinoma this fraction is as high as 90% (Bos, 1989). This oncogenic form of Ras carries mutations (e.g. H-Ras-V12) that decrease the rate of GTP hydrolysis by Ras, causing permanent activation. Thus, oncogenic Ras mimics continuously presentation of growth and survival factors.

MAPK cascade

The cascades downstream of Ras include the well-characterized Raf-Mek-Erk cascade, which is the major subject of this work. Raf binds to the activated

GTP-bound Ras and thereby translocates to the membrane, where it is activated by phosphorylation. Once activated, it can phosphorylate Mek on serines 218 and 222. This double-phosphorylated Mek in turn activates Erk by phosphorylation at threonine 183 and tyrosine 185, with the phosphorylation on tyrosine occurring first (Ferrell and Bhatt, 1997). Activated Erk translocates to the nucleus. Whether this translocation is mediated by an active transport or by diffusion and sequestration in the nucleus is still debated, but it is likely that both mechanisms play a role (Cyert, 2001). Once in the nucleus, Erk activates several transcription factors directly or via nuclear kinases such as Rsk and Msk, and thereby regulates the genetic program. Activated Erk can trigger cell devision, and it is therefore no surprise that Erk is constitutively activated in many cancer cells.

Signal transduction through the same cascade in the same cell type can result in completely different physiological response depending on the kinetic behavior (Kholodenko et al., 1999). For example, in some cell-types such as the neural precursor PC12 the duration of Erk-activation determines the physiological response. In these cells, stimulation with epidermal growth factor (EGF) results in transient Erk-activation that causes proliferation, whereas treatment with nerve growth factor results in sustained Erk-activation that triggers differentiation (O'Neill and Kolch, 2004). It is likely that the interpretation of the signal is carried out in the process of inducing, activating and stabilizing the transcription factors (Murphy et al., 2002). Therefore, it is important to investigate the dynamics after a specific stimulus.

Erk and Mek have two isoforms, Erk-1/2 and Mek-1/2 respectively. In most experiments these isoforms display similar kinetics (e.g. in Woolf et al. (2005)), although in some cell settings Mek 1 and 2 have distinct activation patterns (Xu et al., 1997; English and Sweatt, 1996). However, activation of a single Mek is not coupled to activation of a single Erk isoform, supporting the conclusion that both Mek isoforms phosphorylate both Erk isoforms (Xu et al., 1997). Raf has even more isoforms. Here the activation of the isoforms has different effect on the physiological functions (O'Neill and Kolch, 2004). C-Raf is the most intensively studied of the Raf isoforms. However, there is still controversy and disagreement about precise molecular events that are required for its activation. The activation process is highly complex and involves membrane recruitment, dimerization or oligomerization, binding to other proteins, conformational changes and phosphorylation. Although some of these events seem to be conserved for A-Raf and B-Raf, there are also crucial differences (Wellbrock et al., 2004). A difference exists also in the downstream effectors, and C-Raf signals also to other pathways like NF-κB and to the tumor suppressor retinoblastoma (RB).

In the existing models as well as in this work there is no distinction made between the isoforms of Raf, Mek and Erk. However, when other signal transduction pathways and their crosstalk will be studied, at least the differences in the Raf-isoforms have to be taken into account (Wellbrock et al., 2004).

Transcriptional regulation by Erk

Activation of Erk results in the rapid induction of about 100 genes which have been termed immediate-early genes. Several transcription factors downstream of Erk control these immediate-early genes. These include cAMP response element binding protein (CREB), Elk-1, serum response factor (SRF) (Davis et al., 2000; Rivera et al., 1993), c-Myc, and c-Jun, c-Fos, which form the AP-1 complex (Davis, 1995). However, direct links between these transcription factors and the immediate-early genes are only known anecdotally, and a genome-wide inference of this network is only tried recently by Tullai et al. (2004) using a combination of microarray measurements, bioinformatics and chromatin immunoprecipitation experiments. Although several groups try to gain inside into the links between signal transduction and the induction of genes using clustering of expression data and the search of overrepresented binding sites, these analysis based on pure bioinformatics are still too error-prone (Kielbasa et al., 2004b; Wasserman and Sandelin, 2004).

Objectives of this work

The first part of this thesis is concerned with the activation of Erk. Using dynamical models, a deeper understanding of the processes that mediate Erk-activity is sought. It has been discussed previously that the activation of Erk in *Xenopus* oocytes is rather switch-like, where sub-threshold stimuli are filtered out, but super-threshold stimuli are transmitted causing a strong activation of Erk (Huang and Ferrell, 1996). Such switch-like behavior is termed ultrasensitivity and is the basis for more complicated dynamical properties, such as bistability and oscillations. Huang and Ferrell (1996), the authors of the first mathematical model for this cascade, traced this behavior back to double-phosphorylation of Mek and Erk. As the three-kinase structure of this cascade with double-phosphorylation of Mek and Erk is well preserved from yeast to humans, the question arises, whether also mammalian signal-transduction via Erk is ultrasensitive. This question is addressed from two sides: First the mechanisms that may generate ultrasensitivity in this cascade, namely multisite phosphorylation and enzyme saturation are analyzed with respect to robustness and their properties under high enzyme concentrations as they are present in mammalian cells. Second, the single-cell Erk-activity is quantitatively monitored after hormone stimulation. This way, ultrasensitivity can be quantified and the distribution of activity can be utilized to study the intrinsic variability of ultrasensitivity. Furthermore, these data show that the cascade is not bistable as predicted by the model of Bhalla et al. (2002).

The second part of this thesis investigates the control of the genetic program by Erk. In Chapter 5 the induction of phosphatases is studied as potential signal-terminators. A mathematical model is developed based on time-series data that suggests a role for the phosphatase DUSP6 as signal terminator and tumor suppressor. DUSP6 interferes with Erk-activity and thereby constitutes a negative feedback loop. Then, a more global perspective is taken and the regulation of the genetic program is approached by integrating high-throughput data, sequence data and functional annotation. First, high-throughput experiments are analyzed with respect to the biological processes regulated by Ras and Erk. Second, *in silico* predictions of the biological functions that are regulated by Erk-activated transcription factors are shown. Both approaches require novel methods to handle and interpret gene groups, these methods are developed and discussed.

In the next sections the background of the two parts of the thesis is briefly introduced.

1.2 Models of Ras-mediated signal transduction

For about a decade, Ras-mediated signal transduction, and especially the MAPK-cascades (such the Raf/Mek/Erk cascade downstream of Ras) has been subject to mathematical modeling. Starting with the pioneering work of Huang and Ferrell (1996), several groups have built models describing the quantitative behavior of Erk activation using ordinary differential equations. Vayttaden et al. (2004) have constructed a "evolutionary tree" for these models (see Figure 1.2), showing that the majority of current models are derived from this early work. In the following, the most important models displayed in Figure 1.2 will be briefly introduced.

The model of Huang and Ferrell (1996) was the first published mathematical model of the MAPK-cascade. It was used to analyze the intrinsic ultrasensitivity of the cascade's stimulus-response curve. Although it was constructed for *Xenopus* oocytes, it stimulated modeling affords by many groups that were interested in different organisms. The structure and the

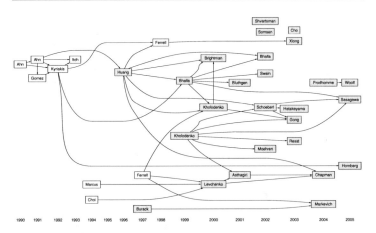

Figure 1.2: The evolution of models that describe the MAPK-cascade (adopted and extended from Vayttaden et al. (2004)). The models are represented by the nodes and named after the first author of the paper that describes the model. Arrows show the information that was used to construct other models. Grey nodes indicate quantitative models, white nodes qualitative models.

parameters were used in several other models for different organisms, such as the model by Bhalla and Iyengar (1999) and Schoeberl et al. (2002) (see Fig. 1.2). In this thesis, this first model will be analyzed with respect to robustness in Chapter 2. Further theoretical investigation of Ferrell and coworkers targeted bistability and hysteresis in the MAPK-cascade in *Xenopus* oocytes (Xiong and Ferrell, 2003; Ferrell and Xiong, 2001).

Modeling by Kholodenko and coworkers was first focused on the activation of adaptor molecules upstream of Ras by the epidermal growth factor receptor (Kholodenko et al., 1999; Moehren et al., 2002), a model which was further extended to the MAPK-cascade and to include receptor internalization by Schoeberl et al. (2002). Lately, these models were further refined. Gong and Zhao (2003) showed that the adapter protein Shc plays a central but redundant role in the signal transduction, and Resat et al. (2003) modeled trafficking of the EGF-receptor. Kholodenko also investigated the appearance of oscillation due to negative feedback (Kholodenko, 2000), which will be discussed in this thesis in the context of substrate sequestration due to high enzyme concentrations in Chapter 3. Recently Kholodenko and coworkers showed that double-phosphorylation in the MAPK-cascade can result in bistability even without considering an explicit positive feedback loop (Markevich et al., 2004).

Bhalla and Iyengar investigated the effect of a positive feedback loop on the dynamics of Erk after stimulation with platelet derived growth factors in mammalian cells as a model for cellular memory. First, using a mathematical model they showed that this feedback might induce bistability in the system, such that sub-threshold stimuli have no significant effect on Erk-activity while super-threshold stimuli cause an irreversible activation of Erk (Bhalla and Iyengar, 1999). Later they showed that Erk-activation is prolonged after short stimulation with PDGF in mouse fibroblasts which is in agreement with their theoretical prediction (Bhalla et al., 2002). They also showed that inhibition of the positive feedback via PLA_2, arachidonic acid and PKC changes the time course in a way that is well in agreement with the experiments. They conclude from the data and the model that the deactivation of Erk after about an hour is performed by the induction of MKP-1 (DUSP1) that shifts the system into a monostable regime. These findings stimulated the investigation of the Erk-activity in single-cells presented in Chapter 4, where we shall show that this pathway is ultrasensitive but not bistable, and that this ultrasensitivity seems to be tightly controlled.

The group of Lauffenburger investigated the dynamics of Erk due to negative feedback regulation (Asthagiri and Lauffenburger, 2001) and due to autocrine loops (Shvartsman et al., 2002). They recently started to use also multivariate statistical modeling to understand the network (Prudhomme et al., 2004; Woolf et al., 2005; Sachs et al., 2005).

Recently, also a model involving cross-talk to another pathway appeared. Hatakeyama et al. (2003) investigated the dynamics of Erk after stimulation of the ErbB4-receptor, and constructed a model involving cross-talk to the Akt-pathway.

Lately, Sasagawa et al. (2005) developed a mathematical model to describe the difference in the dynamics of Erk-activation in PC12-cells, where stimulation with nerve growth factor and epidermal growth factor, result in sustained and transient activation of Erk, respectively. They conclude from their studies that the difference results from the activation of Rap1 by the nerve growth factor receptor. Also they found that the amplitude by which Erk is activated depends rather on the change of the growth factor concentrations than on the absolute growth factor concentration, a finding which has also been reported by Schoeberl et al. (2002). Cho et al. (2003) investigated the effect of RKIP inhibiting Raf-Mek-Erk signal transduction. They showed that this stoichiometric inhibitor RKIP acts nonlinearly on the activation of Erk, and it modulates the amplitude and duration of Erk activity. They conclude that RKIP might have a role in the differential activation of Erk due to nerve and epidermal growth factor.

The MAPK-cascade is probably the best understood mammalian signal transduction cascade, and a variety of specific inhibitors exist for the reactions in this cascade that makes quantitative perturbation experiments possible. Therefore, Hornberg et al. (2005b) used this system to verify experimentally theoretical aspects of control in signal transduction cascades proposed by Heinrich et al. (2002). They could show that phosphatases determine the duration of signaling whereas kinases influence the amplitude, being in agreement with the theory.

To conclude, there exist more than a dozen models, which might be categorized in the following three classes. First, there are models that do not model all biochemical mechanisms in detail like Shvartsman et al. (2002); Asthagiri and Lauffenburger (2001); Hornberg et al. (2005b) to understand more general properties of these signal transduction cascade. Second, there are medium-scale detailed mechanistic models that aim to deduce the dynamic consequences of the biochemical mechanism, like ultrasensitivity due to double-phosphorylation in Huang and Ferrell (1996), and the inhibition of RKIP (Cho et al., 2003). And, third, there are large, detailed models that intend to mimic the true biochemical dynamic for a specific system, like Moehren et al. (2002); Schoeberl et al. (2002); Gong and Zhao (2003); Resat et al. (2003). As the third type of model tries to be as precise as possible, their parameters have a distinct biochemical meaning, they are e.g. "binding rates" or "catalytic rates". Therefore these models can in principle be constructed by detailed kinetic measurements (Teusink et al., 2000). In praxis, however, most of the parameters have not been determined and those which have been determined are often *in vitro* measurements, which might differ from *in vivo* rates. Therefore the kinetic parameters are often determined by fitting the model to time series data. A systematic evaluation of these fitting procedure has shown that time series data can only be used to fix a fraction of the parameters (Stelling and Gilles, 2001; Zak et al., 2003). Often the other parameters do not influence the dynamics significantly, or they influence the dynamics only in combination, and can therefore not be estimated separately (Timmer et al., 2004). Whether these parameters are unidentifiable since the biological system is designed to be insensitive against changes of these parameters is still debated (Stelling and Gilles, 2001).

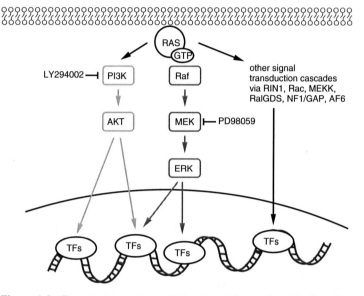

Figure 1.3: Current view on transcriptional modules regulated by Ras. Ras affects several signal transduction cascades, such as the Raf-Mek-Erk cascade and the PI3K pathway. For Mek and PI3K, inhibitors (LY294002 and PD98059) have been used to study pathway specific regulation. Studies of gene expression upon growth-factor stimulation and induction of oncogenic Ras showed that there are groups of genes that are regulated by Erk, by PI3K, by both synergistically, and via other pathways. The figure compiles two figures from Schafer et al. (2003); Campbell et al. (1998).

1.3 Transcriptional regulation by Ras-mediated signal transduction

Upon stimulation of a growth-factor receptor about hundred so called immediate-early genes are rapidly induced via Ras-mediated signal transduction cascades. While the mechanisms of growth-factor driven activation of signal transduction pathways have been analyzed in much detail, the knowledge about the effect of these signals on the genetic program of the cell is rather limited (Campbell et al., 1998; Schafer et al., 2003). Mechanistic details about transcriptional regulation are only known for a few genes, and

| Author | dependent on | | | | |
System	Erk	PI3K	Erk+PI3K	other	total
Tullai et al. (2004) PDGF stimulation	20	11	10	23	74
Zuber et al. (2000) oncogenic H-Ras	↑25 ↓36				244 149[1]
Tchernitsa et al. (2004) oncogenic K-Ras	↑33 ↓46	↑8 ↓0	↑19 ↓0	↑23 ↓52	192
Sers et al. (unpubl.) oncogenic K-Ras	↑113 ↓72				↓274 ↑251

Table 1.2: Number of Ras-responsive genes from different studies. The genes are classified whether they are responsive to Erk or PI3K alone, whether Erk and PI3K regulate the gene synergistically or whether they do not respond to inhibition of these pathways. ↑ and ↓ stand for up- and downregulated genes, respectively.

a global analysis on how signal transduction can yield a specific response is still missing.

Growth-factor induced gene-expression is relayed by a variety of effector cascades downstream of Ras, including the Raf-Mek-Erk cascade and PI3K signal transduction (compare Fig. 1.3). These signal transduction cascades change the activity of a variety of transcription factors, like NF-κB, SRF, Elk-1, AP-1 and FOS. Subsequently, these transcription factors bind to specific transcription factor binding sites. They can interact and form so-called *cis*-regulatory modules. Binding and subsequent interaction with the promoter-bound transcription initiation complexes controls the transcription rate of a gene. When genes share a similar *cis*-regulatory module and are therefore co-regulated, they are often co-expressed, i.e. expressed under the same conditions (Wasserman and Sandelin, 2004). This idea has been reversed by many researchers in recent years. They identify co-expressed genes that are then believed to share common *cis*-regulatory modules.

For Ras-mediated signal transduction this method has been applied to dissect both growth-factor induced transcriptional regulation and regulation due to oncogenic Ras. Tullai et al. (2004) have used inhibitors of Mek and PI3K (Fig. 1.3) in order to identify 74 immediate-early genes whose induction upon growth-factor stimulations depends on the Raf-Mek-Erk cascade, on the PI3K cascade, on both and on none (Tab. 1.2). Then, they computationally

---

[1]ESTs and unknown sequences

identify transcription factor binding sites that might be responsible for this co-expression.

Several studies of genes responsive to oncogenic Ras have been performed by Schäfer, Sers and colleagues. They found that the induction of oncogenic Ras in mammalian cell cultures derived from fibroblast and epithelial cells yields global alterations in gene-expression. Utilizing suppression subtractive hybridization (SSH) they identified 244 differential regulated known genes and 149 expressed sequence tags (ESTs) and unknown sequences (Zuber et al., 2000). About the same number of transcripts are induced and repressed, and the regulated genes are involved in a multitude of processes (Schafer et al., 2003). By using inhibitors they could define modules of Ras-responsive genes.

Whether the gene modules defined by these inhibitor studies are co-regulated by sharing common cis-regulatory modules is still unknown. Computational analysis of these groups by Tullai et al. (2004) and Kielbasa et al. (2004b) lead to different conclusions. The first study found connections to overrepresented binding sites in the promoters of the groups of genes that depend on Erk, PI3K or on both. For some genes this finding could be confirmed experimentally. However, the second study showed that the results of finding these links by pure bioinformatics are dominated by false positives, which confirm the estimate by Wasserman and Sandelin (2004) that one can expect about one functional binding site per thousand predictions. Additionally, the differences in conclusions might also reflect the fact that Tullai et al. (2004) were analyzing immediate-early gene regulation while the gene-groups investigated in Kielbasa et al. (2004b) may include many genes that respond only after sustained Ras-induction. Furthermore, many Ras-responsive genes might be regulated by other means than transcription factors such as DNA-methylation or histone-modification.

Recently, Jürchott et al. (2005) performed a global expression profiling of the effect of signaling inhibitors in several tumor cell lines. This way, cell-system specific responses could be excluded. The data were clustered, and for some clusters an enrichment of binding sites for NF-Y could be shown.

The complexity of signaling networks and transcriptional regulation require a combination of theoretical and experimental methods. Such combined approach is often referred to as systems biology. In the following systems-biological methods will be applied to first investigate ultrasensitivity of Ras-mediated signal transduction, and second to unveil transcriptional modulation by Ras-activation.

Part I

Ultrasensitivity

2

Ultrasensitivity and feedbacks in signal transduction

An extended version of this chapter is to be published in Blüthgen et al. (2006b). The robustness analysis presented in this chapter has been published within Blüthgen and Herzel (2003).

Synopsis

Stimulus-response curves of signal transduction cascades are often non-linear and sigmoidal. The sigmoidality of these curves is referred to as ultrasensitivity to describe the high sensitivity such a system exhibits at low and medium stimuli. This chapter shall review the importance of ultrasensitivity in signal transduction, and especially in the activation of the Raf/Mek/Erk cascade. The major mechanisms that generate ultrasensitivity are introduced. In particular, zero-order kinetics and multisite phosphorylation are discussed as mechanisms that might bring about ultrasensitivity in the Raf/Mek/Erk cascade. These mechanisms are investigated regarding their robustness.

Ultrasensitivity may gives rise to interesting dynamics if signal transduction cascades possess negative and positive feedback loops. The large body of theory for effects of feedbacks shall be reviewed in this chapter. It is discussed that bistability can emerge from ultrasensitivity in conjunction with positive feedback. A cascade with negative feedback can bring about adaptation, oscillations and, surprisingly, highly linear response.

2.1 Introduction

Intracellular signal processing in higher eukaryotes is carried out by signaling networks composed of enzymes that control the activation of each other by covalent modification. Signals at the cellular membrane ripple through these signaling networks by covalent modification events to reach various locations in the cell and ultimately cause cellular responses. The biochemical building blocks of these networks are frequently enzyme couples, such as a kinase and a phosphatase, that form *covalent modification cycles* that can both activate and inactivate a target enzyme by covalently modifying it at single or multiple sites.

The steady-state stimulus-response curves of these curves display strong sigmoidal dependencies *in vivo*, for example in the activation of the MAPK-cascade (Ferrell and Machleder, 1998), in the activation of Sic1 (Nash et al., 2001), and *in vitro*, e.g. in the phosphorylation of isocitrate dehydrogenase (LaPorte and Koshland, 1983), muscle glycolysis (Meinke et al., 1986) and in post-synaptic calcium signaling (Bradshaw et al., 2003). Sigmoidality of the stimulus-response curves has been termed ultrasensitivity, to capture the highly sensitive nature of those systems to changes in signals around a threshold level. In addition, sub-threshold stimuli are dampened, while super-threshold stimuli are transmitted which allows for virtually binary decisions (Ferrell, 1996). Ultrasensitivity can help to filter out noise or can delay responses (Goldbeter, 1991). Moreover it can cause oscillations in cascades composed of interconnected ultrasensitive modification cycles in combination with a negative feedback loop (Kholodenko et al., 2000). Bistability (hysteresis) can occur if such an ultrasensitive cascade is equipped with a positive feedback, as has recently been observed in eukaryotic signaling pathways (Bagowski et al., 2003). Surprisingly, ultrasensitivity can also lead to highly linear signal transduction in the presence of high load, such as translocation to the nucleus (Sauro and Kholodenko, 2004). Mechanisms that lead to ultrasensitive stimulus-response curves include cooperativity, multisite phosphorylation, feed-forward loops, and enzymes operating under saturation. The latter mechanism has been termed zero-order ultrasensitivity because a necessary condition is that both the modifying and the de-modifying enzyme of a covalent modification cycle display zero-order kinetics.

The first part of the thesis investigates ultrasensitivity in the Raf/Mek/Erk cascade. This chapter shall review the theoretical concepts that have been brought up to understand the appearance of ultrasensitivity, bistability, and oscillations. It introduces methods to quantify ultrasensitivity, the means by which ultrasensitivity is generated and how bistability and oscillations arise due to ultrasensitivity and feedback loops.

2.2 Quantification of ultrasensitivity

Metabolic Control Analysis

For analyzing ultrasensitivity, metabolic control analysis (MCA) is a useful tool (Kacser and Burns, 1973; Heinrich and Rapoport, 1974; Fell, 1992). Although developed to study the control of metabolism, it has been successfully applied to intracellular signal transduction systems (Kahn and Westerhoff, 1991; Hornberg et al., 2005b; Kholodenko et al., 1997). MCA links "global" control properties, e.g. the effect of a change in a kinetic parameter (such as a V_{max}) on a flux or a concentration, to "local" properties (e.g. mechanistic details of enzyme reactions). The local properties are called elasticities and are defined by,

$$\epsilon_{x_i}^{v_j} = \frac{[x_i]}{v_j} \frac{\partial v_j}{\partial [x_i]}. \tag{2.1}$$

Elasticities evaluate the relative change in a reaction velocity as a result of an infinitesimal relative change in one of its substrate, product, or effector concentrations (e.g. $[x_i]$). To determine this coefficient, the enzyme is conceptually considered in isolation from the system and only a single metabolite is perturbed; that is, all other metabolites are held constant. The elasticities of an enzyme E_i following irreversible Michaelis-Menten kinetics with the Michaelis-Menten constant K_M are $\epsilon_{E_i}^{v_j} = 1$ with respect to the enzyme concentration and $\epsilon_S^{v_j} = \frac{K_M}{[S]+K_M}$ for the substrate S, respectively.

Global properties are called response coefficients and describe the response of the entire system to small perturbations in parameters (such as rate constants or total concentrations):

$$R_{p_j}^{S_i} = \frac{p_j}{[S_i]} \frac{d[S_i]}{dp_j}. \tag{2.2}$$

Here, $R_{p_j}^{S_i}$ accounts for a relative change in steady-state concentration $[S_i]$ upon a infinitesimal relative change in parameter p_j, therefore $R_{p_j}^{S_i}$ quantifies the relative amplification. Response coefficients higher than 1 correspond to (locally) ultrasensitive systems, where $R < 1$ refers to a sub-sensitive response. Kholodenko et al. (1997) introduced response coefficients for layers of a cascade, where the layers are considered in isolation: $r_i^j = \frac{[K_i]}{[K_j]} \frac{d[K_j]}{d[K_i]}$ is the response coefficient of the activity of kinase j upon changes of kinase i. MCA was designed for the description of steady-state behavior and it is used accordingly in this chapter. However, it was also extended toward transient phenomena and oscillations (Hornberg et al., 2005a,b; Wolf and Heinrich, 2000).

Hill coefficient

In contrast to response coefficients which quantify the sensitivity locally, it is often more meaningful to quantify the sensitivity of the stimulus-response curve over the entire range of stimuli. Such a global characterization is provided by the Hill coefficient, which was first used by Hill as an empirically description of cooperative binding of oxygen to hemoglobin (Hill, 1910). Hill (1910) found that binding was well described by the following relationship, now known as the Hill equation.

$$y = 100 \frac{x^h}{K_{0.5}^h + x^h}, \tag{2.3}$$

where y is bound oxygen in percent of the maximally bound oxygen, x is the oxygen pressure, $K_{0.5}$ is the oxygen pressure where half of the binding sites are occupied, is the maximally bound and h is called Hill coefficient.[1] Cooperative enzymes, such as hemoglobin, have Hill coefficients higher than 1. Hemoglobin has an Hill coefficient of 2.8, and a Hill coefficient of approximately 4 seems to be an upper limit for cooperative enzymes (Cardenas, 1997).

To estimate the Hill coefficient it is common not to fit the formula to the data, but to calculate the Hill coefficient from the cooperativity-index, which is defined as (Goldbeter and Koshland, 1981):

$$R_a = \frac{C_{0.9}}{C_{0.1}}, \tag{2.4}$$

where $C_{0.9}$ is the stimulus-value generating 90% of the response, and $C_{0.1}$ is the value for 10% of the maximum response. To get the relationship between R_a and the Hill coefficient h, the following equations need to be solved:

$$C_{0.9} = 9^{1/h} K_{0.5}, \quad C_{0.1} = K_{0.5}/9^{1/h}. \tag{2.5}$$

Solving for h yields:

$$R_a = \frac{C_{0.9}}{C_{0.1}} = 81^{\frac{1}{h}} \quad \rightarrow \quad h = \frac{\log 81}{\log (C_{0.9}/C_{0.1})}. \tag{2.6}$$

If the Hill coefficient is 1, a 81-fold increase of the stimulus is needed to increase activation from 10% to 90%. For Hill coefficients higher than 1, the

---

[1] Originally, Hill (1910) used the constant differently: $y = 100 \frac{Kx^n}{1+Kx^n}$, thus $K = 1/(K_{0.5})^{1/h}$.

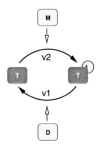

Figure 2.1: Sketch of a simple covalent modification cycle. The modifier protein M catalyzes the modification of the target T, and the demodifying protein D removes the modification from T*. The reaction rates are called v_1 and v_2.

increase of the stimulus needed is smaller and the Hill curve gets sigmoidal. For a Hill function the response coefficient reads:

$$R_a^v = h \left(1 - \frac{a^h}{K_{0.5}^h + a^h} \right) . \tag{2.7}$$

Based on this equation, Legewie et al. (2005a) proposed another method for the quantitative analysis of ultrasensitive systems, which also applies to responses that strongly deviate from the Hill equation.

2.3   Mechanisms

Zero-order ultrasensitivity

Reversible covalent modification of proteins is a frequent regulatory unit in eukaryotic cells regulating nearly all aspects of cellular life (Cohen, 2000). In the beginning of 1980s, Goldbeter and Koshland have shown in a series of articles that a simple cycle of enzymes covalently modifying and demodifying a protein (see Fig. 2.1) can result in highly ultrasensitive responses (Goldbeter and Koshland, 1981; Koshland et al., 1982; Goldbeter and Koshland, 1984). They called this effect zero-order ultrasensitivity since ultrasensitivity in this simple system requires strong saturation of the enzymes, which implies that the reaction velocity is nearly independent of the substrate concentration (called zero-order kinetics). The dynamics of the covalently modified form can be described by one ordinary differential equation for the activated target protein:

$$\frac{d[T^*]}{dt} = v_1 - v_2 , \tag{2.8}$$

where

$$\text{with} \quad v_1 = \frac{V_{max,1}[M][T]}{[T] + K_{m1}} \quad \text{and} \quad v_2 = \frac{V_{max,2}[D][T^*]}{[T^*] + K_{m2}} . \tag{2.9}$$

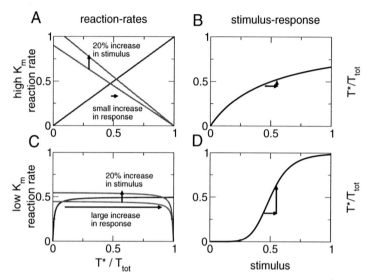

Figure 2.2: The principle of the Goldbeter-Koshland switch. If the enzymes operate under first-order kinetics (A), the stimulus-response curve looks like a Michaelis-Menten curve (B), as opposed to enzymes which are saturated by the substrate (C), where one can observe ultrasensitivity (D).

Here, $[T]$ and $[T^*]$ are the concentrations of the inactive and active form of the substrate, respectively, $[M]$ and $[D]$ are the concentrations of modifier and demodifier. If the enzyme-substrate complexes are negligible (a condition that is discussed in the next chapter), the mass conservation reads

$$[T] + [T^*] = [T_{tot}]. \tag{2.10}$$

Thus, one can express $[T]$ in terms of the total substrate and the active form by $[T] = [T_{tot}] - [T^*]$. In this case the equation for v_1 can be written as

$$v_1 = \frac{V_{max,1}[M]([T_{tot}] - [T^*])}{([T_{tot}] - [T^*]) + K_{m1}}. \tag{2.11}$$

The system is in a steady state if the forward reaction rate v_1 equals the backward reaction rate v_2. This condition can be evaluated graphically, in Fig. 2.2a and 2.2c plots of v_1 and v_2 as a function of $[T^*]$ are shown for two different values of substrate concentrations. In Fig. 2.2a, the substrate concentration is low compared to the K_m values of kinase and phosphatase, whereas in Fig. 2.2c the substrate concentration is high. These plots illustrate how a slight change in the stimulus $[M]$ can either result in a moderate change in response (Fig. 2.2b) or in a drastic change (Fig. 2.2d). Such a covalent modification cycle exhibiting zero-order ultrasensitivity is often referred to as Goldbeter-Koshland switch.

In terms of a response coefficient, zero-order ultrasensitivity can be expressed by (see e.g. Sauro and Kholodenko, 2004):

$$R_{M_T}^{T^*} = \frac{[T]}{\epsilon_{T^*}^{v_2}[T] + \epsilon_T^{v_1}[T^*]}. \tag{2.12}$$

One may consider two cases: the enzymes are not saturated with the substrate (i.e. the elasticities $\epsilon_T^{v_1}$ and $\epsilon_{T^*}^{v_2}$ equal 1), and the case when they are saturated (i.e. $\epsilon_T^{v_1} \ll 1$ and $\epsilon_{T^*}^{v_2} \ll 1$. In the first case, the response coefficient equals $R_{M_T}^{T^*} = \frac{[T]}{[T_{tot}]}$, thus the response coefficient cannot exceed 1. In contrast, if the enzymes are saturated, the denominator becomes small and therefore the response coefficient may exceed 1 and the stimulus response curve may become ultrasensitive.

Theoretical analyzes given in the next chapter demonstrates that zero-order ultrasensitivity is weakened if the enzyme concentration are not negligible when compared to the substrate concentration. Since the concentration of substrate and enzyme are often comparable *in vivo* this causes doubt about the physiological relevance of zero-order ultrasensitivity in signal transduction.

Multiple modification sites

A great fraction of proteins are reversible covalently modified at several sites. As early as in 1976, where only the phosphorylation of five proteins has been studied in detail, it was realized that three of them possess multiple phosphorylation sites (Cohen, 2000). In many proteins the modification is performed by several proteins modifying distinct sites. In this case, multisite phosphorylation can integrate different information sources (Holmberg et al., 2002). For example, activation may require the modification of two sites, controlled by distinct kinases. Then the activity of the protein works as an AND-gate, requiring the presence of both kinases. It may also account for modularity, e.g. to allow the protein to perform several separated functions depending on which stimulus arrives.

In contrast, several proteins such as the epidermic growth factor receptor (EGFR), and the kinases Erk and Mek have multiple phosphorylation sites

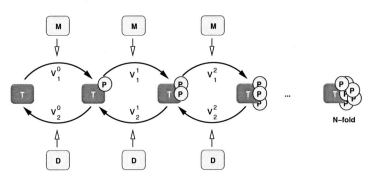

Figure 2.3: Sketch of a multisite modification cycle where the sites are modified by the enzyme M and de-modified by the enzyme D. The index of the target protein T expresses the number of modified sites. This scheme implies that the modification sites are addressed in an ordered manner, e.g. site 1 is the first to be modified, then site two is modified, etc. It is assumed in the calculations that all steps have equal rate constants k_1 for modification and k_2 for de-modification.

that are phosphorylated by the same kinase. It has been widely discussed that the stimulus-response curve of a protein becomes ultrasensitive if activation requires multisite phosphorylation by the same protein (Ferrell and Bhatt, 1997; Deshaies and Ferrell, 2001; Nash et al., 2001; Salazar and Höfer, 2003). In the following this idea is sketched by a simple model for multisite phosphorylation. To distinguish the effect of multisite phosphorylation and zero-order ultrasensitivity, it is assumed that the enzymes are not saturated and are kinetically in their first-order regime, and the enzyme-substrate complexes will be neglected. For simplicity, it is further assumed that the reaction constants for all phosphorylation sites are equal, i.e. the model does not include cooperativity (see below), which is known to enhance ultrasensitivity. Also, it is assumed that the modification sites are modified in an ordered as opposed to a random manner. A reaction scheme for such a system is depicted in Fig. 2.3. The reaction rate of the reaction $T_{i-1} \rightarrow T_i$ is given by $v_1^{i-1} = k_1[M][T_{i-1}]$ and of the reaction $T_i \rightarrow T_{i-1}$ by $v_2^{i-1} = k_2[D][T_i]$. By applying steady-state conditions, one can express the amount of i-fold modified protein by:

$$[T_i] = \left(\frac{k_1[M]}{k_2[D]} \right)^i [T_0]. \qquad (2.13)$$

Assuming that only the full, n-fold modified substrate is active, the normalized steady-state activity is given by

$$\frac{[T_n]}{[T_T]} = \frac{\left(\frac{k_1[M]}{k_2[D]}\right)^n \left(\frac{k_1[M]}{k_2[D]} - 1\right)}{\left(\frac{k_1[M]}{k_2[D]}\right)^{n+1} - 1}, \qquad (2.14)$$

resulting in a response coefficient of

$$R_M^{T_n} = \frac{[M]}{T_n} \frac{dT_n}{d[M]} = \frac{1+n}{1 - \left(\frac{k_1[M]}{k_2[D]}\right)^{n+1}} - \frac{1}{1 - \frac{k_1[M]}{k_2[D]}}. \qquad (2.15)$$

For weak stimulation, the response coefficient equals approximately the number of modification sites n. As the stimulus-response curve of multisite modification deviates significantly from a Hill-curve, the Hill-coefficients are lower than the number of modification sites, e.g. the Hill-coefficients of double-phosphorylation is 1.38 (Blüthgen, 2002). A more detailed study by Kholodenko et al. (1998) on double-phosphorylation includes also saturation effects and yields a response coefficient of

$$R_M^{T_2} = \frac{[T_0](\epsilon_{T_2}^{v_1^1} + \epsilon_{T_2}^{v_2^0}) + [T_1]\epsilon_{T_1}^{v_1^0}}{[T_0]\epsilon_{T_2}^{v_2^1}\epsilon_{T_1}^{v_2^0} + [T_1]\epsilon_{T_0}^{v_1^0}\epsilon_{T_2}^{v_2^0} + [T_2]\epsilon_{T_0}^{v_1^0}\epsilon_{T_1}^{v_2^1}}. \qquad (2.16)$$

This response coefficient is approximately 2 for low stimulation and linear kinetics, but may exceed 2 if the enzymes are saturated. However, the derivation by Kholodenko et al. (1998) did not take high enzyme concentrations into account that are reducing ultrasensitivity (compare next Chapter).

Other mechanisms

The earliest discovered mechanism for an ultrasensitive response was cooperative binding, first discovered in 1904 for the binding of oxygen to hemoglobin (Bohr et al., 1904). Each hemoglobin molecule possesses four binding sites for oxygen. The affinity of a site for oxygen depends on the occupancy of the other binding sites and increases when they are already bound by oxygen. The bound oxygen as a function of oxygen pressure can be reasonably described with a Hill function with coefficient 2.8 (Hill, 1910).

Stoichiometric inhibition and super-sensitization are two mechanisms that can give rise to ultrasensitivity by sequestering the target. Both require high-affinity binding of the target by the stoichiometric inhibitor (Yeung et al., 1999) or by a phosphatase (Legewie et al., 2005b), respectively. The stoichiometric inhibitor or the phosphatase binds the target molecule and prevents it from being active until the target concentration significantly exceeds the phosphatase or the inhibitor.

Another mechanism that generates ultrasensitivity is molecular crowding (Gomez Casati et al., 1999; Aon et al., 2001). This mechanism is based on the finite size of the molecules and needs very high concentrations of the protagonists. Thus it is an unlikely effect in signal transduction, where most molecules are present in only low or medium concentrations.

Sensitivity amplification by a cascade

Brown et al. (1997) pointed out that the response coefficient of a linear signal-transduction cascade is simply the product of the individual response coefficients of each kinase with respect to its upstream kinase:

$$R_1^n = \prod_{i=1}^{n-1} r_i^{i+1} \tag{2.17}$$

This relation simply represents the chain-rule of derivatives. Assuming a three-level cascade where each level responds like a Hill function

$$[K_i] = \frac{[K_{i,tot}] [K_{i-1}]^{h_i}}{[K_{i-1}]^{h_i} + k_{0.5,i}^{h_i}} \tag{2.18}$$

the response coefficient of the terminal kinase K_3 upon a stimulus K_0 reads:

$$R_0^3 = \prod_{i=1}^{3} h_i \times \prod_{i=1}^{3} \left(1 - \frac{[K_{i-1}]^{h_i}}{[K_{i-1}]^{h_i} + k_{0.5}^{h_i}} \right) = \prod_{i=1}^{3} h_i (1 - f_i) \tag{2.19}$$

where f_i is the activated fraction of kinase i. Thus, such cascades can exhibit a sensitivity as high as the product of each kinases Hill-coefficients if they are weakly activated.

2.4 Robustness of multisite phosphorylation and zero-order ultrasensitivity

Variations of parameters not caused by single mutations but by the environment or cell-to-cell variations do not influence single but many parameters. Also allelic variability may change many parameters randomly (Hartman et al., 2001). In the following the effect of such global variability in parameters on ultrasensitivity in two model of the MAPK-cascade (Huang

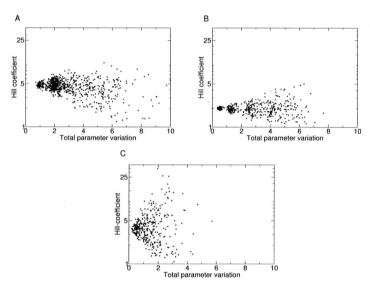

Figure 2.4: Scatter plots showing the influence of random parameter variations for the MAPK-models (A) Huang and Ferrell (1996), (B) Bhalla and Iyengar (1999) and for the Goldbeter-Koshland switch (C) (Goldbeter and Koshland, 1981). T is the total parameter variation $T = \sum_{n=1}^{\#para} |\log \frac{p_i}{p_{i,ref}}|$. Note that the scales are identical to emphasize the pronounced robustness of the cascades in (A) and (B) compared to the Goldbeter-Koshland switch (C)

and Ferrell, 1996; Bhalla and Iyengar, 1999) and in a simple substrate cycle displaying zero-order ultrasensitivity. Such an approach has been applied earlier to the robustness of the bacterial chemotaxis (Barkai and Leibler, 1997). These authors investigated the robustness of the precision and the timescale for adaptation to changes in chemoattractants and -repellents. In analogy to their analysis we generate test parameter sets by multiplying each parameter in the models by 10^x, where x is a random number taken from a Gaussian distribution with mean zero and variance σ. Then we estimate the Hill-coefficient for each test set. Like Barkai and Leibler (1997), we define the "total parameter variation" as

$$T = \sum_{i=1}^{\#para} |\log_{10} \frac{p_i}{p_{i,\text{ref}}}|, \qquad (2.20)$$

where $\#para$ is the number of the parameters in the model and p_i and $p_{i,\text{ref}}$ are the parameters in the reference parameter set and in the test set. The total parameter variation T can be interpreted as the total order of magnitude of parameter variation. Subsequently, we plot the estimated Hill coefficient for each test set as a function of the total parameter variation T in scatterplots (Fig. 2.4). If the total parameter variation as proposed by Barkai and Leibler (1997) is a reasonable measure for distances in kinetic parameter sets, it can be seen that the Hill coefficient of the MAPK-cascade seems to be more robust with respect to random parameter fluctuations than in a covalent modification cycle displaying zero-order ultrasensitivity, the so-called Goldbeter-Koshland switch. To double or halve the Hill coefficient in the MAPK-cascade, a total parameter variation of three orders of magnitude is needed, whereas a total parameter variation of $T < 1$ might be sufficient to change the Hill coefficient in the Goldbeter-Koshland switch by a factor of two. An explanation of this robustness is that the random changes in a large cascade like the MAPK-cascade are more likely to compensate each other, while in a small system like the Goldbeter-Koshland switch compensation is less probable. Moreover, the different mechanisms of generating ultrasensitivity play a role. Since in the MAPK cascade not the saturation but the adjustment of the working-range determines the Hill-coefficient (Blüthgen, 2002), a change of a kinetic parameter in the activating reaction can be compensated by a change of the activity of the corresponding enzyme. In contrast, loss of enzyme saturation in the Goldbeter-Koshland switch cannot be compensated by other parameters and allways yields a reduction in sensitivity.

2.5 Effect of feedbacks on ultrasensitive cascades

There exists a multitude of feedbacks in Ras-mediated signal transduction via the Raf/Mek/Erk cascade. These feedbacks act on all levels of signaling (compare Fig. 2.5). They involve induction of hormones that form autocrine feedbacks, the transcriptional regulation of entities in the cascade, their covalent modification or receptor internalization. In the following the consequences of such feedback loops on the dynamics are briefly reviewed.

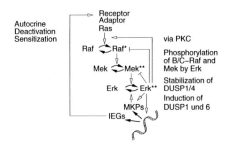

Figure 2.5: Feedbacks in Ras/Raf/Mek/Erk signal transduction can be found at all levels. They involve direct modification, regulation of the phosphatase stability, autocrine signaling, and changes in expression.

Bistability due to positive feedback

Many theoretical and experimental investigations have shown that positive (or double negative) feedback loops in gene regulatory circuits can induce bistability (e.g. Thomas et al., 1976; Arkin et al., 1998; Hasty et al., 2000; Gardner et al., 2000; Becskei et al., 2001). Such a system exhibits two stable steady-states. Often the the coexistence of the steady-states is a function of a stimulus, therefore the system can be switched from one state to another. Also a positive feedback loop wrapped around a signal transduction cascade can cause bistability (Ferrell and Xiong, 2001; Ferrell, 2002; Shvartsman et al., 2002). A prerequisite for observing bistability is that either the cascade or the feedback loop is ultrasensitive (Tyson et al., 2003), as described below.

As discussed above, the processive double-phosphorylation of Mek and Erk suggest that the cascade is ultrasensitive (Chapter 4 shows experiments that suggest that PDGF-induced signal transduction via Raf/Mek/Erk is ultrasensitive.). Therefore, a positive feedback loop might induce bistability also in this signal transduction cascade, which has been confirmed experimentally in *Xenopus* oocytes (Ferrell and Machleder, 1998). Bhalla and colleagues have shown that also in the mammalian Raf/Mek/Erk cascade bistability can arise from a positive feedback loop via PKC, with time series data from mouse fibroblast supporting their hypothesis (Bhalla and Iyengar, 1999; Bhalla et al., 2002). Unlike Ferrell and colleagues they have not shown single-cell data displaying that each cell responds in an all-or-none manner. In Chapter 4 single-cell measurements are shown, that are in contrast to Bhalla's hypothesis, that there is bistability in this system. Nevertheless, there are indications that another MAPK-cascade, the Jnk-cascade is bistable also in mammalian systems (Bagowski et al., 2003).

An intuitive, graphical way to investigate whether a system is bistable is

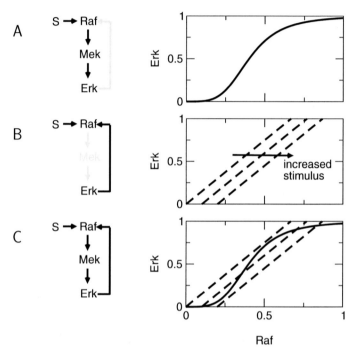

Figure 2.6: How bistability arises: First the feedback is inhibited and the stimulus response curve from Raf to Erk is drawn (A). Second, the cascade is inhibited and the Erk activity is modulated to get the steady-state of Raf. As Raf is dependent on the stimulus S and Erk activity, the curve shifts toward higher Raf activity for higher stimuli (B). The intersecting points of these curves in the combined plot correspond to the steady-states (C). There are three situations: One intersecting point at low Erk-activity (corresponding to one stable off state), three intersecting points (corresponding to a stable off- and on state and an unstable state in between) and an intersecting point at high Erk-activity (a monostable system with high Erk-activity in steady-state). Depending on the slopes of the curves, all three situations might be reached by varying the stimulus. From the plot it is apparent, that at least one curve needs to have a point of inflection to get three intersecting points.

a combined plot of the stimulus response curve and the effect of the feedback loop (Angeli et al., 2004; Bhalla and Iyengar, 1999). This can be done as follows (see also Fig. 2.6): First one blocks the feedback loop and records the steady-state output as a function of a steady-state input. Then one blocks the signal-transduction cascade, perturbs the terminal kinase (the output) and measures the influence on the first kinase. As the curves represent lines where the cascade or the feedback are in steady-state, respectively, their intersecting points are the steady-state for the entire system.[2] In Figure 2.6 such curves are displayed for different stimuli. From these curves it becomes clear that at least one of the curves must have a point of inflection and thus must be ultrasensitive to have three intersecting points.

If both the stimulus-response-curve and the feedback loop are not ultrasensitive the system possesses maximally two steady-states, one unstable and one stable. For networks containing feedback circuits with non-ultrasensitive stimulus response curves, Binder and Heinrich (2004) and Heinrich et al. (2002) have investigated the conditions under which the ground state is stable or looses its stability. If the network possesses a stable ground state, the system shows only transient activation. Otherwise the system would exhibit permanent activation, which might contribute to tumor-genesis (Heinrich et al., 2002).

Linear response, adaptation, and oscillation due to negative feedback

Whereas positive feedback tends to destabilize the stable "off" state and often creates a second stable steady state, the existence of a negative feedback loop usually leads to a stabilization of the steady state (Sauro and Kholodenko, 2004; Becskei and Serrano, 2000; Heinrich et al., 2002). Three emergent dynamic phenomena have been described for ultrasensitive signal-transduction cascades and negative feedback loops: a linearization of the response (Sauro and Kholodenko, 2004), damped oscillations (interpreted as adaptation, see Asthagiri and Lauffenburger, 2001; Blüthgen and Herzel, 2001) and sustained oscillations (Kholodenko, 2000; Blüthgen, 2002). In the following these three phenomena will be briefly reviewed.

---

[2]This plot corresponds to null-clines in the phase-plane for a two dimensional system. The conclusions drawn from these null-clines in higher dimensions than two are only valid in case both stimulus-response curves for the feedback and signal-transduction cascade are monotonic, see Angeli et al. (2004) for details. Moreover, the conclusions do not necessarily hold for cascades with multiple feedback loops.

Linear response

Sauro and Kholodenko (2004) were the first who related the stimulus response curve of ultrasensitive cascades with those of an operational amplifier, a device often used in analog electronic circuits together with a negative feedback to get a linear response. They showed that a negative feedback leads to a linear response over a wide range of stimuli. Kholodenko et al. (1997) derived the response coefficient of a cascade with a feedback loop:

$$R_1^T = \frac{R_{\text{cascade}}}{1 - r_T^1 R_{\text{cascade}}} . \qquad (2.21)$$

Here R_{cascade} is the response coefficient of the cascade in isolation, r_T^1 is the local response coefficient of the first kinase in the cascade upon changes in the targets (i.e. the effect of the feedback loop in isolation). In case of a linear negative feedback loop, $r_T^1 = -1$ and thus R_S^T simplifies to:

$$R_1^T = \frac{R_{\text{cascade}}}{1 + R_{\text{cascade}}} , \qquad (2.22)$$

if the first kinase is not saturated by the stimulus. Surprisingly, the response becomes linear if the response coefficient of the cascade becomes large compared to 1. This result depends strongly on the assumption that $r_T^1 = -1$. Otherwise the response is approximately:

$$R_1^T \approx -r_T^1 . \qquad (2.23)$$

Therefore such a negative feedback loop together with a highly ultrasensitive cascade can be a strategy to "outsource" control of the sensitivity to the feedback. As long as the sensitivity of the cascade is high, only the sensitivity of the feedback determines the sensitivity of the system. Such a strategy might lead to higher robustness if the feedback is rather simple, such as Erk modifying a upstream molecule. Then, only this reaction has to be tightly controlled to yield an amplifier with robust sensitivity.

Adaptation

Often only the information that some concentration has changed is important, whereas information about the absolute value is not needed. For example, rising concentrations of growth factors indicate wounding and neighboring cells need to respond by migration and proliferation. If the signal is prolonged, however, this behavior might lead to cancer, and thus the signal has to be terminated after some time. Another example it the sensing of nutritional gradients in bacterial chemotaxis (Alon et al., 1999), where the bacteria need to respond to changes only as their size does not permit to sense a gradient directly. To transduce only the information that something changes, the signal-transduction cascades need to adapt, i.e. become less sensitive to higher stimuli when a prolonged stimulus is given, and regain sensitivity if the stimulus drops. Lauffenburger (2000) distinguishes two types of adaptation: perfect and partial adaptation. Perfectly adapting systems show transient activation but have a steady-state output that is insensitive to the signal. In contrast, partially adapting systems show a peak of activity after stimulation but reach a steady-state that is higher than before stimulation.

A common motif in signal-transduction to gain adaptation is the negative feedback loop (Lauffenburger and Linderman, 1993). A negative feedback will introduce damped oscillations in the cascade (see Appendix A). If a stimulation is given to the system the target protein will be activated after some delay and will cause e.g. the de-sensitization of the receptor, which leads to cascade adaptation. Many Ras-activating receptors and adapter molecules are desensitized this way. That a simple negative feedback system, where the terminal kinase desensitizes the receptor cannot perform perfect adaptation is already apparent from Eqn. 2.21: To perform perfect adaptation, the steady-state needs to be insensitive to the stimulus, i.e. $R_1^T = 0$. Thus the cascade itself has to be insensitive to the stimulus, i.e. $R_{cascade} = 0$, but this implies that the cascade itself has to perform adaptation. This is also intuitively clear, as if the steady-state of the target kinase is insensitive to the stimulus, it cannot provide any information about the stimulus strength that can be used to desensitize the receptor. Therefore other mechanisms have to be exploited, if perfect adaptation is required. One possibility is to feed the integrated output into the system, e.g. to have a molecule that has a very slow dynamics and integrates the difference between the desired steady-state output and the actual output, as it is realized in bacterial chemotaxis (Yi et al., 2000). However, perfect adaptation is probably not required in all systems, since e.g. weak steady-state activation after partial adaptation of Erk may be insufficient to activate downstream targets.

Oscillations

If the cascade or the feedback loop are strongly ultrasensitive, the oscillations might be sustained. In Kholodenko (2000) it was shown that a three-layer cascade can exhibit sustained oscillations, if the overall sensitivity exceeds a threshold value that is determined by the timescales in the cascade. In Appendix A this relation is derived and it is shown that oscillations appear if the overall sensitivity (the sensitivity of the cascade and the feedback) exceeds 8, provided that the timescales of all levels of the cascade are equal. Although sensitivity of eight is unlikely in the Raf/Mek/Erk cascade (compare Chapters 3, 4), sustained oscillations have been reported to be observed experimentally (Sauro and Kholodenko, 2004), but they have not been published. In other systems with negative feedback, sustained oscillations have physiologically important roles, such as in circadian rhythms, cell cycle and development.

2.6 Discussion

The Raf/Mek/Erk signal transduction cascade has two mechanisms that may result in an ultrasensitive response: Mek and Erk are processively double-phosphorylated and as each kinase is controlled by phosphorylation-dephoshphorylation cycles they might exhibit ultrasensitivity due to enzyme saturation. Moreover, as the cascade is composed of the three layers, Raf, Mek and Erk, ultrasensitivity generated by each layer amplifies along the cascade. In Chapter 4 it is shown that the cascade is indeed ultrasensitive with a Hill coefficient of above 2.

The question arises why cells need such a complicated mechanism like double-phosphorylation to gain ultrasensitivity, while a simple phosphorylation-dephosphorylation cycle may give rise to a similar degree of ultrasensitivity. One idea is that each layer may mediate cross talk to other cascades. Another argument was given in this chapter: Such a cascade provides a much more robust way to generate ultrasensitivity. Additionally it will be discussed in the next chapter that sequestration due to similar enzyme/target concentrations limits zero-order ultrasensitivity.

Ultrasensitivity is of importance to noise filtering: Low stimuli which may occur by chance are filtered out. Additionally, feedback loops in conjunction with ultrasensitivity can generate interesting dynamics. For example, ultrasensitivity is needed to build molecular switches, i.e. to generate bistability together with positive feedbacks. Ultrasensitivity together with negative feedback is a strategy to gain robust amplifiers. Here, a highly ultrasensitive cascade "outsources" the control over sensitivity to the properties of the feedback, so that only one point needs to be controlled. It is, however, of importance that ultrasensitivity is kept in a certain range, otherwise the system might show sustained oscillations. Therefore it is of importance for the cell to have mechanisms like multisite phosphorylation that can tolerate parameter fluctuations without drastically changing the degree of ultrasensitivity. Partial adaptation can also be caused by negative feedbacks through damped oscillations, a phenomenon discussed in Chapter 5.

3

Sequestration limits ultrasensitivity

This chapter is the result of a collaboration with Frank Bruggemann, Boris N. Kholodenko, and Herbert M. Sauro. It is an extended version of Blüthgen et al. (2005b).

Synopsis

The building blocks of most signal-transduction pathways are couples of enzymes, such as kinases and phosphatases, that control the activity of protein targets by covalent modification. Goldbeter and Koshland (1981) have demonstrated that these systems can be highly sensitive to changes in stimuli if their catalyzing enzymes are saturated with their target protein substrates. This mechanism, termed zero-order ultrasensitivity, may set thresholds that filter out sub-threshold stimuli. Experimental data on protein abundance suggests that the enzymes and their target proteins are present in comparable concentrations. Under these conditions a large fraction of the target protein may be sequestered by the enzymes. This causes a reduction in ultrasensitivity so that the proposed mechanism is unlikely to account for ultrasensitivity under conditions present in most signaling cascades *in vivo*. Furthermore, we show that sequestration changes the dynamics of a covalent modification cycle and may account for signal termination and a sign-sensitive delay. Finally, we analyze the effect of sequestration on the dynamics of a complex signal transduction cascade: the MAPK-cascade with negative feedback. We show that sequestration limits ultrasensitivity in this cascade and may thereby abolish the potential for oscillations induced by a negative feedback.

3.1 Introduction

Ultrasensitivity in the Raf/Mek/Erk cascade might be generated by at least two mechanisms: double-phosphorylation of Mek and Erk, or zero-order ultrasensitivity, i.e. enzymes operating under saturation (Huang and Ferrell, 1996). The latter mechanism was explored for the steady-states of cycles composed of enzymes with irreversible product-insensitive kinetics in the pioneering work by Goldbeter and Koshland (1981). It is very appealing because of its simplicity: all it needs is one modification site on a protein that acts as substrate (for example a phosphorylation site) and modifying enzymes that each have a K_M value for their protein substrate that is low compared to the concentration of that substrate. However, the experimental evidence showing that high sensitivity actually results from this mechanism *in vivo* is lacking. In fact, cells seem to possess more complicated mechanisms to bring about ultrasensitivity such as the ones that activate proteins by multiple modification events. Examples of such protein targets are Sic1 which has at least 6 phosphorylation sites that all need to be phosphorylated before it shows activity (Nash et al., 2001). NFAT has even more phosphorylation sites (Salazar and Höfer, 2003). Also the MAPK-cascade Raf/Mek/Erk that is subject of this thesis contains Mek and Erk that both become fully activated by double phosphorylation. It is still a puzzle, why other, more complicated means like multisite phosphorylation need to be applied to get high sensitivity when there is a simple mechanism like zero-order ultrasensitivity.

Moreover the conditions derived by Goldbeter and Koshland (1981) have been scrutinized by Ortega et al. (2002) who show that as the enzymes are product sensitive, ultrasensitivity is unlikely to obtain. Numerical simulations showed that other mechanism as proposed by Goldbeter and Koshland (1981) lead to higher robustness with respect to parameter fluctuations (see previous chapter). In this chapter another argument is added: sequestration due to high enzyme concentrations affects zero-order ultrasensitivity. Additionally, it is shown that sequestration has strong effects on the dynamics of signal transduction cascades. First, sequestration allows for signal termination without involving negative feedback. Second, sequestration and multisite phosphorylation can cause sign-sensitive delay of signal transduction. Third, as sequestration reduces zero-order ultrasensitivity it dramatically compromising the potential of the MAPK-cascade to generate oscillations.

Cell type	MAPKKK	MAPKK	MAPK	Ref.
Budding Yeast		$< 35nM$	$100nM$	Ferrell (1996)
CHO cells		$1300nM$	$2800nM$	Ferrell (1996)
Xen. oocytes	$3nM$	$1200nM$	$330nM$	Ferrell (1996)
HeLa cells		$30\mu M$	$30\mu M$	Schoeberl et al. (2002)
Rat 1	1 rel.u.	1.6 rel.u.	2.4 rel.u.	Yeung et al. (1999)
NIH 3T3	1 rel.u.	1.4 rel.u.	3.5 rel.u.	Yeung et al. (1999)
208F	1 rel.u.	2.9 rel.u.	5.9 rel.u.	Yeung et al. (1999)
COS-1	1 rel.u.	0.7 rel.u.	9 rel.u.	Yeung et al. (1999)

Table 3.1: Concentrations of members of the MAPK-cascade (MAPKKK, MAPKK, MAPK) in different organisms and cell-types as found in the literature. In many of these, the concentrations are of the same order of magnitude.

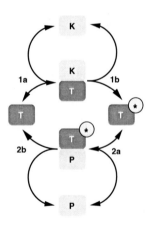

Figure 3.1: Schematic representation of the simplest covalent modification cycle. The target protein T can be (de-) modified covalently. The unmodified protein T binds to the kinase K in the first reaction ($1a$) to form the complex KT. The second reaction ($1b$) is the catalytic modification step yielding K and the covalently modified protein T^*. In the third reaction ($2a$), the demodifier P binds T^* to form the complex T^*P. In the fourth reaction ($2b$) the cycle is closed by the recycling of T through catalytic demodification and the release of P. The reactions $1b$ and $2b$ are assumed to be irreversible.

3.2 Methods

This chapter applies methods from metabolic control analysis, which has been introduced in the previous chapter. In the first part of this chapter a simple covalent modification cycle is analyzed that consists of two enzymes K and P that modify and de-modify a target protein T, respectively (see Fig. 3.1). T can be in a modified and un-modified form, denoted by T^* and T, espectively. To investigate the effect of sequestration, we model the reactions catalyzed by the two enzymes K and P where enzyme-target complex are explicitly taken into account. For the case of product sensitivity, this system has been shown not to be ultrasensitive, and therefore such effects are not considered here (Ortega et al., 2002). Ortega et al. (2002) did non consider sequestration and high enzyme concentrations. Salazar and Hofer (2006) showed by analytical calculations that ultrasensitivity in covalent modification cycles for product-sensitive enzymes with high concentrations can only be obtained for some special cases. Therefore, we neglect product-sensitivity in the following. The total concentrations of the three enzymes involved are denoted by $[T_T]$, $[K_T]$ and $[P_T]$. The dynamics of this kinetic scheme depicted in Fig. 3.1 is described by a system of three ordinary differential equations using mass-action kinetics. Additionally, the effect of sequestration in a more complicated system, the MAPK-cascade, is analyzed. The details of the kinetic model can be found in the Appendix B. The numerical analysis of the equations used Mathematica and xpp-auto Ermentrout (2002).

3.3   Sequestration in covalent modification cycles

Unlike metabolic systems, the modification cycles involved in signal transduction cascades often exhibit comparable amounts of protein substrates and enzyme (Albe et al., 1990). For example, the individual concentrations of the three kinases of the well characterized MAPK-cascade are similar in a variety of cell types and organisms (Table 3.1). Each of these kinases modifies its target protein and is itself a target for the upstream kinase. Since the concentration of kinases and their target proteins are comparable, the kinase can sequester a significant amount of target by binding to it provided that the kinase shows a high affinity toward the substrate. This sequestered fraction of the target is no longer accessible to other kinases and phosphatases. Available data about phosphatase concentrations suggests that phosphatase concentrations are also likely to be in the same order of magnitude or even exceed their substrate concentrations (Brown and Kholodenko, 1999; Goldberg, 1999).

The concentration of the kinase-substrate complex $[TK]$ in steady-state can be calculated by the Michaelis-Menten formula:

$$[TK] = \frac{[T][K_T]}{[T] + K_M},\tag{3.1}$$

with [T] and $[K_T]$ being the free target concentration and the total kinase concentration, respectively, and K_M being the Michaelis-Menten constant of the kinase. The concentration of the complex $[TK]$ approaches the total concentration of the kinase as $[T]$ becomes larger than K_M. The phosphatase-substrate concentration can be calculated accordingly. To illustrate this effect for a covalent modification cycle, we investigate a special case, i.e. when both kinase and phosphatase have the same kinetic constants and the same concentrations. Consequently, the two substrates are equal in their steady-state concentration ($[T] = [T^*]$). Also the two complexes equal ($[TK] = [T^*P]$). Therefore, the total target concentration can be expressed as: $T_T = 2[T] + 2[TK]$. After substitution of the resulting expression for $[TK]$ into the Michaelis-Menten formula, we obtain:

$$2\frac{[K_T][T]}{K_M + [T]} = T_T - 2[T].\qquad(3.2)$$

From this, the amount of free substrate in the cycle, i.e. $[T] + [T^*] = 2[T]$ can be calculated from the total concentrations of kinase and target. Importantly, the concentrations of the free substrate forms $[T]$ and $[T^*]$ drop below the K_M-values if $[K_T] > [T_T] - 2K_M$ (see appendix for mathematical details). If the catalytic activity of the phosphatase exceeds the activity of the kinase, the free substrates can be higher. In this case, $[T]$ and $[T^*]$ will still drop below the K_M-values if the kinase and phosphatase concentration together exceed twice the target concentration. This seems to be the case in vivo, as measured phosphatase concentrations largely exceed their target concentrations.

In a signaling cycle sequestration reduces the free target concentrations such that the concentration of the free target is below the K_M-value of the enzymes, provided that the enzymes are available in a concentration as high as their total protein substrate concentration.

3.4 The effect of sequestration on zero-order ultrasensitivity

The sensitivity of simple modification cycles has been explored in pioneering work by Goldbeter and Koshland (1981) using methods from nonlinear dynamics. Later, it has been formulated in terms of metabolic control analysis (MCA) by Small and Fell (1990). Small & Fell expressed the response of the active fraction to a change of the kinase concentration as a function of the concentrations of the two forms ($[T]$ and $[T^*]$) and the elasticities of the enzymes by the following simple relation:

$$R_{K_T}^{T^*} = \frac{[T]}{\epsilon_{T^*}^{v_2}[T] + \epsilon_T^{v_1}[T^*]}.\qquad(3.3)$$

As discussed in the previous chapter, this response coefficient expresses the fractional change of the active form $[T^*]$ upon a fractional change of the kinase concentration. If the enzymes are unsaturated, their elasticities are $\epsilon_{T^*}^{v_2} \approx 1$ and $\epsilon_T^{v_1} \approx 1$, and the response coefficient is below 1, corresponding to a sub-linear response. In this case no ultrasensitivity is observed. In contrast, saturation of the enzymes leads to elasticities closer to zero, hence $R_{K_T}^{T^*}$ can exceed 1 and give rise to an ultrasensitive response. In the derivation of Eq. 3.3, Small and Fell (1990) assumed that the concentration of the substrate bound to the enzyme is negligible. But as discussed above, this assumption does not hold where the concentrations of enzymes and substrate are similar, as observed in signal transduction cascades if the enzymes are saturated. Therefore, the assumptions made to derive Eq. 3.3 may not necessarily hold.

If the effect of sequestration is taken into account the response coefficient modifies to:

$$R_{K_T}^{T^*} = \frac{[T]}{\epsilon_{T^*}^{v_2}[T] + \epsilon_T^{v_1}[T^*] + \epsilon_{T^*}^{v_2}\epsilon_T^{v_1}\left([TK] + [T^*P]\right)} \cdot \qquad (3.4)$$

A detailed mathematical derivation of Eqn. 3.4 can be found in the appendix. Comparison of Eq. 3.4 with Eq. 3.3 reveals the effect of sequestration on zero-order ultrasensitivity as an additional term in the denominator which increases with the extent of sequestration, i.e. $([TK]+[T^*P])$. Therefore, at constant elasticities, sensitivity should decrease with sequestration. Another effect is hidden in the equations: An increase in sequestration also increases the elasticities $\epsilon_{T^*}^{v_2}$ and $\epsilon_T^{v_1}$, since the available substrate decreases. This eventually causes an additional decrease of the sensitivity $R_{K_T}^{T^*}$. To elucidate this further, we examined the special case when both kinase and phosphatase have the same kinetic constants as we did previously. In this case, we expect on the basis of symmetry that the highest response coefficient occurs when there is an equal amount of phosphorylated and unphosphorylated target. We can then express all concentrations and elasticities in terms of $[T]$, the Michaelis-Menten constant K_M of kinase and phosphatase. In this case Eq. 3.4 reads:

$$R_{K_T}^{T^*} = \frac{1 + \frac{[T]}{K_M}}{2\left(1 + \frac{K_M[K_T]}{(K_M+[T])^2}\right)} \qquad (3.5)$$

$R_{K_T}^{T^*}$ increases with $[T]$ and decreases with $[K_T]$. This shows that the response coefficient gets smaller as the amount of free substrate, $[T]$, decreases due to sequestration. As discussed before, similar concentrations of the enzymes and taget imply that the free target falls below the K_M value. Then the response is sublinear, i.e. $R_{K_T}^{T^*} < 1$, since $\frac{2K_M}{[T]+K_M} \approx 1$. Additionally, if the

K_M is very small, most of the substrate is sequestered, leading to essentially zero concentration of T and T^*.

Goldbeter & Koshland discussed the possibility that ultrasensitivity might be preserved if the phosphatase-target complex T^*P is assumed to be active. However, as calculated in the appendix, the combined response of T and T^*P, $R_{K_T}^{T^*+T^*P}$, is always below $R_{K_T}^{T^*}$. Thus the attenuation of sensitivity by sequestration cannot be restored by an active phosphatase-target complex.

3.5 The consequences of sequestration for ultrasensitivity: numerical investigations

To investigate further the consequences of sequestration on ultrasensitivity, the steady state of the cycle depicted in Fig. 3.1 was calculated numerically. The K_M value was chosen to be much smaller than the total concentration ($K_M = 0.02[T_T]$) for both the kinase and the phosphatase. The phosphatase concentration $[P_T]$ was increased from 0 to $2[T_T]$, to vary the amount of sequestration. Fig. 3.2B, demonstrates that this increase is accompanied by an increase in the sequestered fraction $([TK] + [T^*P])/[T_T]$. The response of the cycle $[T^*]$ to the input $[K_T]$ decreases if the total levels of the phosphatase approaches half of the total target concentration $[T_T]$ (cf. Fig. 3.2A). Taken together these two plots illustrate our argument: when the kinase and phosphatase concentrations become comparable to the total concentration of the target protein, the sequestered fraction increases, which causes the sensitivity to *decrease*. In Fig. 3.2C the activated fraction of the target T^* is plotted, illustrating that the fraction of activated target drops dramatically as the phosphatase concentration exceeds $[T_T]/2$. These results are well in agreement with the estimates made in the previous section. This suggests that *in vivo*, where in many cases the concentrations of the kinase, the phosphatase, and the target protein are comparable, sensitivity of covalent-modification cycles is likely to be achieved by other mechanisms than zero-order ultrasensitivity. Simulations for different catalytic activities of kinases and phosphatases are shown in Fig. 3.2D-I. If the phosphatase is catalytically ten-fold more active than the kinase, the region of enhanced sensitivity is slightly broadened (Fig. 3.2D-F). In contrast, if the kinase is more active than the phosphatase, the region if ultrasensitivity is drastically reduced (Fig. 3.2G-I). It seems that for the MAPK-cascade, where the phosphatases have higher activity than the kinases, that the high concentrations of the phosphatases put the system in a regime where sequestration allows for no ultrasensitivity.

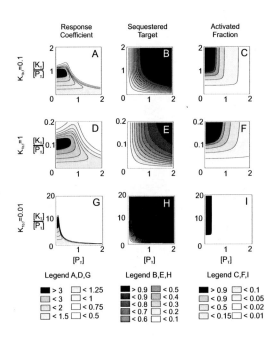

Figure 3.2: Steady-state signaling characteristics of a covalent-modification cycle for equal catalytic activity of kinase and phosphatase (A-C), for ten-fold higher catalytic activity of the kinase (D-F), for ten-fold reduced catalytic activity of the kinase (G-I). A: a contour plot of the response coefficient $R_{K_T}^{T^*}$ as function of the total concentration of the phosphatase and the kinase (normalized to the phosphatase concentration), B: the sequestered fraction of the target protein, C: the fraction of the activated target protein. Parameter values: $T_T = 1$, $k_{1a,f} = 10$, $k_{1a,r} = 0.1$, $k_{1b,r} = 0$, $k_{2a,f} = 10$, $k_{2a,r} = 0.1$, $k_{2b,f} = 0.1$, and $k_{2b,r} = 0$. $k_{1b,f}$ was varied to simulate different catalytic activity of the kinase: A-C: $k_{1b,f} = 0.1$, D-F: $k_{1b,f} = 1$, G-I: $k_{1b,f} = 0.01$. K_T and P_T refer to the total kinase and phosphatase concentration, respectively.

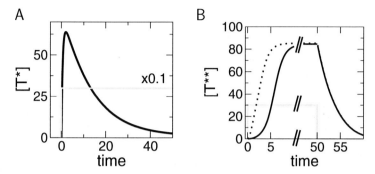

Figure 3.3: A: The dynamics of free phosphorylated target protein in case of more active kinase than phosphatases ($k_{1a,f} = 0.005$, $k_{1a,r} = 0.4$, $k_{1b,f} = 0.1$, $k_{2a,f} = 0.0005$, $k_{2a,r} = 0.004$, $k_{2b,f} = 0.001$ $T_T = 100$ $K_T = 300$ $P_T = 300$). At zero time-point, the system is at steady-state for zero stimulus (initial conditions: $[T](0) = T_T$, $[T^*](0) = [TK](0) = [T^*P](0) = 0$). B: The dynamics of double-phosphorylation in case the kinase shows higher affinity towards the unphosphorylated target. Solid line: the case of kinase sequestration, dotted line: no kinase sequestration. Gray lines indecate the stimulus (i.e. kinase concentration), scaled by 0.1 in A.

3.6 The consequences of sequestration for signaling dynamics

Receptor desensitization is a relatively slow process and downstream signal transduction cascade are often in quasi steady-state with the receptor activity. However, some downstream parameters adapt very quickly (e.g., IRS-phosphorylation after Insulin and Erk after EGF), suggesting that downstream pathways are capable of adaptation. Figure 3.3A displays the dynamics of the covalent modification cycle for a fast kinase with low affinity and a slow phosphatase with high affinity. If a permanent stimulus is given, the target displays only transient activation. Thus a covalent modification cycle is capable of terminating prolonged signals. The fast kinase phosphorylates the available target, but the phosphorylated target gets subsequently sequestered by the low-activity high-affinity phosphatase. At steady-state most of the target substrate is sequestered by the phosphatase. Thus substrate sequestration by a phosphatase might be a means to achieve signal termination and desensitization downstream of receptors without the need of a negative feedback loop.

For many signals, their duration determines the biological response (Marshall, 1995). We pointed out that sequestration might cause short, transient signals. However, the interpretation of the signal by the signal transduction network requires circuits that respond only to prolonged activation. As pointed out by Deshaies and Ferrell (2001), such signal duration decoding requires a threshold time. Additionally, deactivation must be fast in comparison to activation as removal of the signal has to be translated into an immediate response. Such properties have been described for coherent feed-forward loops, which display sign-sensitive delay (Mangan et al., 2003). Fig. 3.3B shows that competition for the enzyme by two phosphorylation sites may also account for such a sign-sensitive delay and dramatically improves duration decoding. The solid line shows the dynamics of double-phosphorylation where both phosphorylation sites compete for the kinase, the dotted line shows the dynamics of the corresponding system in case there is no competition (details in the appendix). If the stimulus rises, the stimulus has to be of a certain length to be transduced if the sites compete for the kinase. However, if the stimulus falls, the change is immediately transduced. Thus, sequestration and multisite-phosphorylation might be a mechanism for sign-sensitive delays, similarly to coherent feed-forward loops in transcriptional networks (Mangan et al., 2003).

Additionally, changes in the steady-state stimulus-response curve might also have a large impact on the dynamics because the onset of oscillations in a signal transduction cascade harbouring a negative feedback is determined by the sensitivity of the stimulus-response curve in steady state (see previous chapter, at least a response coefficient of 8 is required). We have investigated the effects of sequestration in a complex signal transduction cascade with negative feedback as described in the next section.

3.7 The effect of sequestration in MAPK signal transduction cascade

The MAPK-cascade consists of three kinases that activate their downstream kinases by phosphorylation. It has the potential to be highly ultrasensitive due to the combination of multisite-phosphorylation, zero-order kinetics (Huang and Ferrell, 1996), and cascade amplification effects (Brown et al., 1997). According to Kholodenko (Kholodenko et al., 2000) a negative feedback that is wrapped around this ultrasensitive cascade can bring about sustained oscillations over a wide range of stimuli if sequestration is neglected

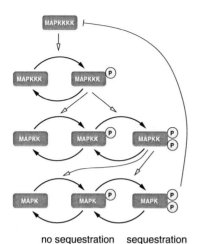

Figure 3.4: Sketch of the MAPK cascade. A MAPKKKK stimulates the phosphorylation of MAP-KKK, which after phosphorylation phosphorylates MAPKK at two sites. The double-phosphorylated MAPKK phosphorylates MAPK also at two sites. The double-phosphorylated MAPK in turn inhibits the activity of MAPKKKK.

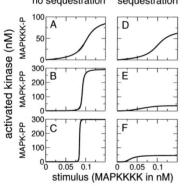

Figure 3.5: Stimulus-response curves for the three layers of the MAPK-cascade in the model considered by Kholodenko (Kholodenko et al., 2000) (A-C), which neglects sequestration and the corresponding model that takes the effects of sequestration into account (D-F).

(Fig. 3.6A).

As the kinases are present in similar concentrations, we investigate whether sequestration effects ultrasensitivity and oscillatory behavior. We modeled the cascade such that sequestration was taken into account (see appendix), and chose the concentrations of the phosphatases for MAPK and MAPKK to be as high as that of their substrate (300 nM). First we analyzed the cascade without a feedback. Figure 3.5 compares the model that neglects sequestration (A-C) and the model that includes the effect of sequestration (D-E). While the response of the first molecule (MAPKKK) is

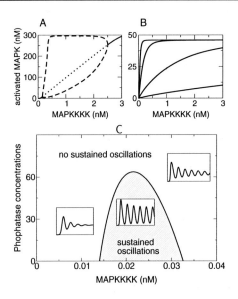

Figure 3.6: A and B: Bifurcation diagrams for the models that neglects (A) and that includes the effects of sequestration (B+C). Solid lines show stable steady-states, dotted lines indicate unstable steady-states. The dashed lines mark the amplitude of the oscillations observed in the model that neglects sequestration. The four lines in graph B show situations for different feedback parameters (from top to bottom: k_{loop}=9, 1, 0.1, 0.01nM). C: Two-dimensional bifurcation diagram for the model that includes the effect of sequestration. Concentrations of the MAPK- and MAPKK-phosphatases (vertical axis) and the stimulus (horizontal axis) are changed. The dashed area shows the region where sustained oscillations occur. Insets show qualitatively the dynamics in the corresponding areas.

relatively unchanged since its kinase and phosphatase are only present at low concentrations, the response of the second and third molecule (MAPKK and MAPK) is changed dramatically. There are mainly two effects of sequestration visible in the response of MAPKK and MAPK: the ultrasensitivity of the stimulus response curves is reduced and the amount of maximally activated kinases in this cascade is decreased.

If we add a negative feedback loop to this model similar to the model by

Kholodenko et al. (2000), no oscillations arise (see Fig. 3.6B). The effect of lower activation of MAPK can be compensated by a stronger feedback (lower values of k_{loop}, see appendix). However, lowering of k_{loop} does not restore oscillations (see Fig. 3.6B). This leads us to conclude that the reduction in ultrasensitivity due to sequestration is responsible for the diminishing of oscillations.

We observed in the analysis of simple, isolated covalent modification cycles that an increase of the total target concentration will limit the sequestered fraction of the target and restore ultrasensitivity. However, in cascades such as the MAPK-cascade the kinases are both enzymes for the modification of the downstream kinase and substrate for the upstream kinase. Hence, the complex of e.g. MAPK and MAPKK reduces the free concentration of both MAPK and MAPKK. Therefore an increase of the MAPK concentration in this cascade gives rise to more sequestration of MAPKK by MAPK. Consequently it is not surprising that we found that an increase of MAPK by one order of magnitude cannot restore the oscillations.

In addition, we investigated the effects of sequestration by phosphatases. We found that oscillations can be restored if the phosphatase concentrations of MAPK- and MAPKK-phosphatase are lowered to a fifth of the kinase concentrations (while increasing their catalytic activity by factor five to keep the V_{max} constant). However, in contrast to the model that neglects sequestration, the stimulus has to be rather low (see Fig. 3.6C). In this case the sequestration due to the phosphatase is reduced and the upstream kinases of MAPK and MAPKK are only slightly activated and can sequester only limited fractions of MAPK and MAPKK.

3.8 Discussion

The function of the signal transduction network is to sense changes in the environment of the organism in the form of signals of physico-chemical origin, e.g. concentrations of molecules, mechanical stress, and integrate these with the current cellular status to 'compute' an adaptive response (Sauro and Kholodenko, 2004). Such adaptive responses involve covalent modification of enzymes, changes of gene expression, and cell-fate decisions that occur on different time scales. Many signal transduction networks have common building blocks, which are enzyme couples that activate and inactive their protein targets by way of covalent modification. It is reasonable to expect that network responses can be highly sensitive to changes in the signals. Ultrasensitivity can be used generate thresholds, oscillations, and linear responses (Sauro and Kholodenko, 2004). Therefore, it may not be surprising that ultrasensitivity has been documented experimentally for some signaling systems (Bagowski et al., 2003).

Theoretical studies by Goldbeter & Koshland (Goldbeter and Koshland, 1981) unveiled a potential mechanism responsible for ultrasensitivity: the kinase and phosphatase have to be saturated with their target protein. This case was referred to as zero-order ultrasensitivity. Since then, many groups have analyzed zero-order ultrasensitivity, e.g. see Sauro (1994); Cardenas and Cornish-Bowden (1989); Ortega et al. (2002). Although the effect of complex-formation in substrate cycle has been addressed before by (Fell and Sauro, 1990), the impact of sequestration on zero-order ultrasensitivity has not been addressed.

Experimental data (Table 1) indicate that the concentrations of enzymes and target proteins in signal transduction cascades are similar. When the affinity of enzymes for their target is sufficiently high, it implies that a high fraction of the target concentration is bound to the enzymes, and thereby sequestered. This, in turn, decreases the amount of target accessible to the enzymes, and reduces ultrasensitivity. Moreover, the amount of the activated target decreases dramatically. Consequently, the concentrations of the complexes can no longer be neglected in the analysis of ultrasensitivity, as long as the concentrations of players in the signal transduction cascades are comparable.

In this chapter we investigated the consequences of sequestration on zero-order ultrasensitivity using analytical approach of metabolic control analysis (MCA) and numerical simulations. In terms of MCA, ultrasensitivity is equivalent to a response coefficient higher than one (Ortega et al., 2002). We derived an analytical expression for the response coefficient (Eqn. 3.4) that accounts for the effect of sequestration. A comparison with the response coefficient that neglects sequestration (Eqn. 3.3) suggests that sequestration may significantly reduce and even eliminate ultrasensitivity. Eqn. 3.5 corroborates this for a simple example where the kinetic parameters of both enzymes are equal. It shows that the response coefficient drops below 0.5: hence, ultrasensitivity is absent. The results of numerical simulations illustrated that if the total concentrations of both enzymes are increased simultaneously, ultrasensitivity decreases and ultimately vanishes when these concentrations exceed 70% of the total target concentration. This correlated with a high sequestration of the target protein by the enzymes, which illustrates that sequestration reduces ultrasensitivity.

Another problem of zero-order ultrasensitivity arises due to the sequestration of the enzyme by the substrate: The saturated enzyme may then not be available for other reactions. This is of special importance if the enzyme itself is a substrate of a modification cycle like MAPKK which is itself controlled by phosphorylation and is the enzyme that phosphorylates MAPK. Here sequestration reduces the zero-order ultrasensitivity in both cycles: the cycle where the enzyme drives the modification and the cycle where the enzyme is subject to modification. In such signaling cascades sequestration can be significant even if the kinases concentrations increase along the cascades due to the sequestration of the enzymes. The extent of ultrasensitivity that can be generated by signal transduction cascades is thereby limited by sequestration. This effect might be responsible for the fact that sustained oscillations have not yet been documented in the MAPK cascade as opposed to the NF-κB cascade (Nelson et al., 2004). Since usually each enzyme targets more than one reaction, as for example most phosphatases, modification cycles compete for the enzymes. After a pathway is activated it recruits its phosphatases, which are no longer accessible to others. We show that sequestration of the kinase in a double-phosphorylation cycle may account for a delay element, such that the activation of a target enzyme upon a signal is delayed, but it is in-activated immediately after removal of the signal. Such a delay element provides cells with units that neglect short fluctuations in signals, but transduces long signals.

Additionally, sequestration might mediate cross-talk between pathways if an enzyme is shared. This has been observed in the JAK/STAT pathway, where the receptors share the janus-kinase (JAK) and multiple receptors compete for it. An up-regulation of one receptor down-regulates the response of the other by sequestration of JAK (Dondi et al., 2001).

Our results suggest that for generating ultrasensitivity, cells need to exploit mechanisms that do not require enzyme saturation. Such mechanisms include multisite-phosphorylation, which generates ultrasensitivity without the necessity of sequestration. Moreover, not only ultrasensitivity, but also bistability and hysteresis arise from multi-site covalent modification in signaling cascades (Markevich et al., 2004). Ultrasensitivity and bistability induced by multi-site phosphorylation may be a widespread mechanism for the control of protein activity in signaling networks, whereas zero-order ultrasensitivity is unlikely to be the major means of generating switch-like behavior in such systems.

One thing is clear, the covalent cycle is extremely versatile for eliciting different kinds of behavior (Sauro and Kholodenko, 2004; Legewie et al., 2005b; Salazar and Hofer, 2006). This great versatility may partly explain why signalling pathways, in both prokaryotic to eukaryotic systems, employ this motif in so many instances. Unfortunately the lack of any clear guidance from experimental data means we are are unable to determine exactly the true functional role played by these motifs. Although many signalling networks have been mapped out in great detail we still have very little understanding of their actual dynamical behavior. Until experimentalists embrace a systems approach we will remain in the dark regarding this question.

4

Erk-activation upon PDGF stimulation is ultrasensitive

Anja Scharamme and Jana Keil performed the stimulation of the cell-lines. Details on quantification of the immuno-flourescence images can be found in Appendix C.

Synopsis

The preceding chapters have highlighted that the Raf/Mek/Erk signal transduction cascade has the potential to be ultrasensitive since Mek and Erk require double phosphorylation to become activated. Ultrasensitivity may give rise to bistability, if the cascade possesses a positive feedback loop. In an interplay between modeling and experiments, Bhalla et al. (2002) have shown that a positive feedback loop via cPLA$_2$ and PKC may cause a prolonged Erk activation upon PDGF stimulation. They concluded that this behavior is due to bistability in the system. In this chapter, the distribution of Erk-activation in single cells is investigated. First, consequences of cell-to-cell variations and stochasticity are investigated in a model according to Bhalla et al. (2002), and the results are compared to single-cell measurements of Erk-activity in mouse fibroblasts stimulated with PDGF. The signaling cascade turns out to be ultrasensitive with a Hill-coefficient of about 4, but not to be bistable. Moreover, the distribution of Erk-activity suggests that the threshold varies with a standard deviation of about 15%, which is lower than measured fluctuation in gene expression.

4.1   Introduction

Signal transduction cascades may exhibit two stable steady-states, if they are ultrasensitive and possess a positive feedback loop (compare Chapter 2). If the stimulus exceeds a certain threshold, the low-activity steady-state can lose its stability and the system approaches a high-activity steady-state. Depending on the parameters, the high-activity steady-state can remain stable even if no stimulus is present. Thus the cascade stays activated until some external parameter changes the stability of this state. Such irreversible transitions have interesting biological interpretation: They can be associated with check-points. A stimulus has to exceed a threshold to start a process, but once it is started the process is maintained. Such a behavior might also be biologically advantageous in the Raf/Mek/Erk cascade, as this cascade controls processes that require irreversible decisions such as cell-devision, survival and differentiation.

As Mek and Erk are controlled by double-phosphorylation, the Raf/Mek/Erk cascade is likely to be ultrasensitive, and a positive feedback loop can bring about bistability (as discussed in Chapter 2). Indeed, in *Xenopus* oocytes, the activity of $p42_{MAPK}$ (homologue to mammalian Erk), has been shown to be bistable due to a positive feedback loop (Ferrell and Machleder, 1998; Xiong and Ferrell, 2003). Also the Jnk MAPK-cascade was shown to be bistable in *Xenopus* oocytes (Bagowski and Ferrell, 2001). In contrast to the orthologous mammalian Jnk cascade (Bagowski et al., 2003), there is no direct evidence that Erk-activation is bistable in mammalian cells. It has been shown recently by single-cell measurements that Erk activation due to EGF is not bistable in mammalian cells (Whitehurst et al., 2004). However, whereas EGF stimulation transiently elevates Erk-activity, PDGF stimulation yields sustained Erk-activity (Murphy et al., 2002). Bhalla et al. (2002) hypothesized that the prolonged Erk-activity is caused by a positive feedback loop. They have developed a model for the Raf/Mek/Erk cascade in mouse fibroblasts that includes a positive feedback loop via $cPLA_2$ and PKC and a negative feedback loop via DUSP1 (MKP1). They showed that this model can reproduce time series of Erk-activation after stimulation of PDGF. In particular, the model could well explain why Erk-activation is prolonged after a short stimulation (5 min) with PDGF. The rationale is that the activation of Erk is bistable until DUSP1 is induced that destabilizes the high-activity steady-state. Thus, Erk response is prolonged even after brief stimuli, since signal termination requires elevated levels of DUSP1. Additionally, the model predicted well the effect of inhibitors against PKC and $cPLA_2$, and the effect of elevated DUSP1-levels. In the light of the work by Whitehurst et al. (2004), this feedback needs to act PDGF-dependent, as

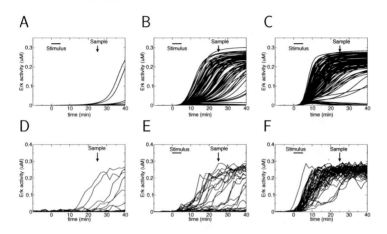

Figure 4.1: A,B,C: Time series of 100 deterministic simulations with varying initial conditions for PDGF stimuli of 0.0001 nM, 0.2 nM, 1 nM. D,E,F: Time series of 50 stochastic simulation for PDGF stimulations of 0 nM, 0.05 nM and 0.5 nM. The interval during which the stimulation with PDGF was performed is indicated by a horizontal bar, the time point at which Erk-activity was sampled to generate the histograms (Fig. 4.2) are indicated by an arrow.

EGF-stimulated cells show only brief Erk-activation. Indeed, Hornberg et al. (2004) have shown that the feedback via PKC and cPLA$_2$ does not influence the dynamics of Erk-activation upon EGF-stimulation.

This chapter discusses that bistable activation of Erk will bring about all-or-none activation at the single cell level. It shows predictions of the model by Bhalla et al. (2002) for Erk-activity in a population of heterogeneous cells and compares these results with data obtained in 3T3 mouse fibroblasts. It turns out that Erk-activation, although prolonged, does not show all-or-none response at the single cell level. Thus Erk-activity is not bistable upon PDGF stimulation. Additionally the data shows that Erk activation by PDGF is ultrasensitive, and that the threshold at which half of the maximal activation is achieved seems to be tightly controlled.

A   B

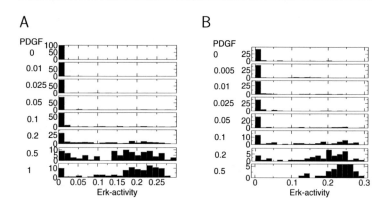

Figure 4.2: Histograms of activity of Erk 20 min after 5 min stimulation with PDFG (nM) as predicted by the model according to Bhalla et al. (2002), A: stochastic simulations in a cell volume of $1.6 \times 10^{-18} m^3$, B: varying initial concentrations with standard deviation of 2.5 on the logarithmic scale (base 10).

4.2   Heterogeneity in a population of cells

Even cloned cells of a cell line show heterogeneity in their shape, size, and behavior. Also signal transduction may be effected by heterogeneity in a population. First, the constituents of signal-transduction cascades are present in a number that is determined by stochastic processes like gene expression. Thus the number of molecules for each species differs between cells (Thattai and van Oudenaarden, 2004; Pedraza and van Oudenaarden, 2005; Rosenfeld et al., 2005). Second, as the numbers of molecules for each species might be small, the reactions constituting the network might of stochastic nature (Bhalla, 2004; Shibata and Fujimoto, 2005). Both effects should not have large impact if the system is far from a bifurcation point. However, in a bistable system, the activation of a few molecules by chance might lead to a "jump" from low activated Erk to high activity.

In this section, it is investigated how heterogeneity in a population of cells caused by stochastic gene expression and stochasticity in the reaction shapes the distribution of Erk-activity. The model proposed by Bhalla et al. (2002) will be used to simulate the influence of stochastic effects on Erk-activity in a population of cells, using the following protocol. The model is first allowed to relax to the steady-state. Subsequently, a stimulation with PDGF of varying concentrations for 5 min is simulated. 20 min after stimulation, Erk-activity is recorded. As on the time scale of 20 min regulation of DUSP1 had no effect on Erk-activity in the original model, DUPS1 dynamics was not included in this simulation to keep computation feasible.

Varying concentrations

To model the effect of different concentrations of the constituents of the cascade among cells, the model was simulated 100 times with varying initial concentrations. Each initial concentration from the original model $C_i(0)$ was multiplied by a random number from a logarithmic Gaussian distribution:[1]

$$\tilde{C}_i(0) = C_i(0) \times 10^\xi \qquad (4.1)$$

Time-series for three different stimuli are shown in Figure 4.1A-C. After stimulation, some cells change from no activity to high Erk activity, while others only show transient and very low activation. Higher stimuli result in more cells that switch to higher activity. Additionally, the time scale of the transient from low to high activity differs substantially from cell to cell, and higher stimuli enhance the transition.

The time course of Erk-activity cannot be observed experimentally in single cells. However, a snap-shot of the distribution of Erk-activity can be obtained by *in-situ* immuno-histochemistry, and can be compared to the simulation. The distribution of Erk-activity 20 minutes in the simulations is shown in Fig. 4.2A. The simulations predict a bi-modal distribution with a peak at low activity and a peak at high-activity. Interestingly, even at high stimuli some cells show no significant activation of Erk since the system becomes mono-stable due to the variations in concentrations.

Stochastic reactions

Heterogeneity among cells might also be caused by stochastic nature of the reactions. Although fibroblast cells are relatively large, the reactions might happen in smaller compartments. Therefore, in the compartments the number of involved molecules might be low. To study stochastic effects in the reaction the model was reformulated as master equation, and simulated using

---

[1]$\tilde{C}_i(0)$ is the initial concentration i with $C_i(0)$ being the original value published by Bhalla et al. (2002). ξ is drawn from Gaussian distribution with zero mean and standard deviation of $\sigma = 0.025$. In addition to $\sigma = 0.025$, other values were tried ($\sigma = 0.05, 0.075, 0.1, 0.125$) yielding the same conclusions.

the Gibson-Bruck algorithm (Gibson and Bruck, 2000).[2] Time-series of the stochastic simulations are displayed in Fig. 4.1D-F. Also here, cells respond in an all-or-none manner. However, even without stimulation, some cells switch from the off-state to the on-state (Fig. 4.1D). Additionally, all cells can be elevated to high-activity upon strong stimulation, thus stochasticity in the reactions does not destroy bistability. The distribution of Erk-activity at 20 minutes after stimulation from 50 stochastic simulations are shown in Fig. 4.2B.

There are two major differences in the distributions obtained by simulations with varying concentrations and stochastic reactions: First, having heterogeneous concentrations of signaling molecules in the population does not allow cells to have Erk-activity without stimulation. In contrast, stochasticity in the reaction gives rise to (very few) cells in the snapshot that display Erk-activity devoid stimulation. Second, varying concentrations of molecules among cells give rise to a number of cells that lose bistability, e.g. that have only one stable state. Therefore, in Fig. 4.2A there are several cells that exhibit no Erk activity even under high stimulation.

A common feature of both is that Erk-activation is bi-modal for stimuli that are around the threshold. In a bistable system, each cell responds in an all-or-none manner. However, cell-to-cell variations lead to a distribution of thresholds in the population such that mean activation of the population is graded. This gives rise to a bi-modal distribution where the peak at low-activity shrinks and the peak at high activation rises as the stimulus increases. This phenomenon has been discussed in depth for transcriptional networks (Gardner et al., 2000; Biggar and Crabtree, 2001; Becskei et al., 2001; Hofer et al., 2002), and it has been also appreciated for signal-transduction cascades by Ferrell and colleagues (Ferrell, 1998; Bagowski and Ferrell, 2001; Ferrell, 2002; Bagowski et al., 2003).

In the following the results from simulation will be compared to measurements of Erk-activity in single cells.

4.3 Erk activity in mouse fibroblasts

The last section has shown that bistability of Erk-activation implies a bi-modal distribution of Erk-activity in a population of cells. In this section, the

---

[2]The Gibson-Bruck algorithm is equivalent to the Gillespie-algorithm but has a better performance for large sparse networks (Gillespie, 1977). The algorithm was implemented in C++, and is available at http://sourceforge.net/projects/pathsim/, the results are shown for a cell volume of $1.6 \times 10^{-18} m^3$. Additional simulations were performed for a cell size of $1.6 \times 10^{-17} m^3$, and resulted in similar distributions.

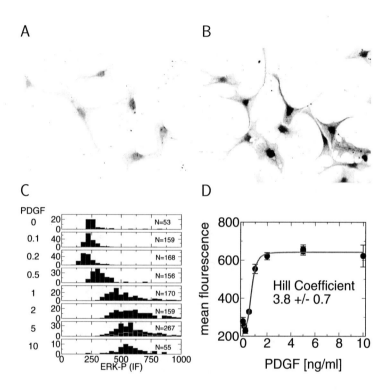

Figure 4.3: A,B: Distribution of phosphorylated Erk in single cells without (A) and with stimulation by 10 ng/ml PDGF (B). C,D: Histograms (C) and mean value (D) of Erk-activity measured by immuno-flourescence. Error bars indicate standard deviation of the mean. The curve is the maximum likelihood fit using a Hill equation. The fitted curve shows a Hill coefficient of 3.8±0.7.

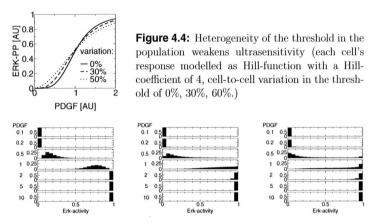

Figure 4.4: Heterogeneity of the threshold in the population weakens ultrasensitivity (each cell's response modelled as Hill-function with a Hill-coefficient of 4, cell-to-cell variation in the threshold of 0%, 30%, 60%.)

Figure 4.5: Prediction of activation histograms with different variances of the threshold (15%, 30% and 50%)

prediction is tested experimentally. The experiment is performed in serum-starved 3T3-cells. Equivalently to the protocol simulated above, the cells are stimulated with PDGF of varying concentrations for 5 min, washed and fixated after 20 min. Subsequently, they were immuno-stained with anti-pErk antibodies and flourescently labeled. Using fluorescent microscopy, dozens of cells per stimulus were imaged, and their fluorescence in the nucleus was quantified (see Fig. 4.3A,B for sample images, the details of the quantification and normalization of the images are discussed in Appendix C). The distribution of Erk-activity clearly shows a mono-modal distribution that is shifted to higher values for higher stimuli (compare Fig. 4.3C). This mono-modal distribution is in contrast to the prediction for a bistable signal-transduction cascade, thus Erk-activity is not bistable.

Plotting the mean values of Erk-activity for each stimulus, the maximum likelihood fit with a Hill curve yields a ultrasensitive curve with a Hill coefficient of 3.8 (Fig. 4.3D). Assuming that the fluorescence readout of Erk-activity increases linearly with the concentration of phosphorylated Erk, this finding suggests ultrasensitive activation of Erk by PDGF. As the data used in Fig. 4.3D to estimate ultrasensitivity is the mean over the population, it is be investigated in the following how ultrasensitivity of single cells is reflected in the mean of population.

Heterogeneity in 3T3 cell-lines

First, the influence of heterogeneity in a population on the mean stimulus-response curve is analyzed. For simplicity, only variation in the threshold, i.e. the stimulus required to achieve 50% maximal activation is considered, although also maximal activation and ultrasensitivity might vary from cell to cell. The mean Erk-activation might be estimated by:

$$\langle Erk(PDGF)\rangle = \int_0^\infty p(k)\frac{\left(\frac{PDGF}{k}\right)^h}{1+\left(\frac{PDGF}{k}\right)^h}dk. \tag{4.2}$$

Here, p(k) is the distribution of thresholds in the population, which is modeled with truncated Gaussian noise,

$$p(k) = \frac{\frac{1}{\sigma\sqrt{2\pi}}e^{\left(\frac{k-\bar{k}}{2\sigma}\right)^2}}{1-\int_{-\infty}^0\frac{1}{\sigma\sqrt{2\pi}}e^{\left(\frac{k-\bar{k}}{2\sigma}\right)^2}dk}. \tag{4.3}$$

The resulting stimulus-response curves for the mean of the population are displayed in 4.3D. As a consequence of heterogeneity in the population the stimulus-response curve of the mean is less sensitive than that found in individual cells. Thus, the Hill coefficient of Erk-activation in single cells is underestimated in Fig. 4.3D, and may exceed 4.

Also, the distribution of Erk activity is shaped by cell-to-cell variation of the threshold. Especially, if the stimulus is around the mean threshold of the cells variations of the threshold might lead to a broad distribution. The distribution can be calculated by

$$p(Erk|PDGF) = p(k(Erk))\frac{1}{\left|\frac{dH\left(\frac{PDGF}{k}\right)}{dk}\right|}, \tag{4.4}$$

with

$$H\left(\frac{PDGF}{k}\right) = \frac{PDGF^4}{k^4+PDGF^4} \tag{4.5}$$

being the Hill-function of degree 4, and

$$k(Erk) = H^{-1}(Erk, PDGF) = \sqrt[4]{\frac{PDGF^4}{Erk}-PDGF^4}. \tag{4.6}$$

being the inverse of the Hill function to estimate the threshold for a given Erk and PDGF concentration. Resulting histograms are displayed in Fig. 4.5. As the variation of the threshold increases, the resulting histograms become broader if the stimulus is around the mean threshold. Additionally, for high variation of the threshold the peak of the distribution is at all or no activation. In contrast, small variation of the threshold ($\sigma \approx 15\%$) gives rise to a distribution observed experimentally: As the stimulus approaches the threshold, the distribution moves towards higher activation.

4.4 Discussion

Bistability gives rise to all-or-none responses on the single-cell level, as cells switch rather rapidly from an off-state to an on-state after the stimulus exceeds a threshold. Heterogeneity in the population will lead to different thresholds in each cell. Therefore, for a given stimulus some cells will respond, while others stay in their off-state. This gives rise to a bimodal distribution of activity, while the mean activity over the population rises gradually. Thus, the all-or-none response can only be measured in single-cell experiments, and Western blots will not show all-or-none activation.

Quantification of Erk-activity in single cells show that Erk-activity is mono-modal and rises gradually upon PDGF-stimulation, which suggests that Erk-activation is not bistable. Data by MacKeigan et al. (2005) confirmed this result. Analysis of their data yields a Hill-coefficient above 5 (see Appendix C). Consequently, the question arises how Erk-activity is maintained over about half an hour after stimulation with PDGF, since it drops quickly after stimulation with EGF. One possibility is that the differences are due to different rates of receptor deactivation.

The experimental data show that mean Erk-activation is ultrasensitive upon PDGF-stimulation with a Hill-coefficient of about 4. It has been reported previously, that activation of the PDGF receptor is ultrasensitive with an Hill coefficient of 1.7 (Park et al., 2003). As ultrasensitivity quantified by the Hill coefficient is at most the product of the Hill coefficient of subsequent layers in the cascade, the cascade from receptor to nucleus has to be ultrasensitive with a Hill coefficient of at least 2.2.

Due to stochasticity in gene expression and the resulting heterogeneity in the population, the threshold stimulus required by individual cells to display 50% of maximal activation varies from cell to cell. As a consequence, the Hill-coefficient of mean activation is reduced when compared to stimulus-response curves of single cells. Thus, the Hill-coefficient might be even higher than 2.2. The cell-to-cell variation of the threshold are estimated by comparing theoretically derived histograms of activation and those experimentally found. The comparison suggests that the threshold varies with a standard deviation in the order of 15%. The noise in gene expression was quantified in yeast and E. coli to be in the order of 30% and 60%, respectively (Raser and O'Shea, 2004). Most contribution to the noise in expression were found not to be intrinsic due to the transcriptional and translational machinery, but to exist mainly due to extrinsic factors, such as the activity of transcription factors or variations of the general capacity of the cells to generate proteins (Colman-Lerner et al., 2005). As the expression of phosphatases determines the threshold mainly linearly, it is surprising that the cells investigated here show such small variations regarding their threshold. Most probably, gene expression in mammalian cells is much more tightly controlled than in lower eukaryotes or prokaryotes.

The results of this chapter show that the role of the feedback loop via PKC and cPLA$_2$ is not understood. One possible role of the feedback loop via PKC and cPLA$_2$ might be in enhancing the cascade's ultrasensitivity. In further studies, one may block the this feedback loop as in Bhalla et al. (2002) and Hornberg et al. (2004) and investigate single-cell activity. Additionally, a DUSP1 (MKP1) knock-down might be interesting as the cells investigated in this chapter might display elevated DUSP1-levels when compared to the cells in Bhalla et al. (2002) although they are the same cell types and the experiments follow the same protocol.

Part II

Transcriptional regulation

5

Negative feedback regulation by phosphatases

The microarray data used in this chapter is generated by Oleg Tchernitsa, normalized and clustered by Ralf-Jürgen Kuban, and the Western/Northern blots have been made by Jana Keil.

Synopsis

In the preceding chapter the activation of Erk due to growth factor stimulation was discussed. This chapter shall discuss the dynamics of Erk activation due to oncogenic Ras. Some mutations in Ras results in a permanent, growth factor-independent activation of the downstream signaling pathways including the Raf-Mek-Erk cascade. Through this, they cause the induction and repression of various genes. Since oncogenic Ras signals receptor-independently, it circumvents several negative feedback loops that target receptors or adaptors and usually terminate the signal (Tsang and Dawid, 2004; Langlois et al., 1995). It is clear that growth-factor induced Erk activation is terminated by adaptor-targeted feedback. However, the mechanism for signal termination of growth-factor independent activation of Erk such as by oncogenic Ras, calcium or phorbol 12-myristate 13-acetate (PMA) induced signal transduction is not known. Interestingly, activated Erk induces a variety of dual-specific phosphatases (DUSPs), including DUSP1, DUSP2, DUSP5, and DUSP6[1] which dephosphorylate Erk and thereby form negative feedback loops. The physiological role of the induction of phosphatases is

---

[1]DUSP1, DUSP4, DUSP5, and DUSP6 are also known as MKP-1, MKP-2, cpg21 and MKP-3, respectively

A   B

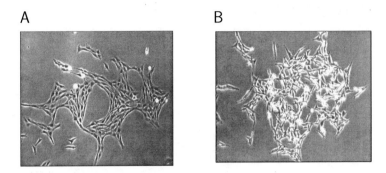

Figure 5.1: Morphological changes after the induction of oncogenic Ras in IR4-cells. Cells before (A) and 48h after (B) the induction of H-Ras V12 (Figure taken from Sers et al. (2002)).

unclear, and it is not known which phosphatase is important on which time scale. Previous theoretical studies have shown that such negative feedback loops can mediate adaptation, therefore it is investigated here whether the induction of DUSPs is a mechanism of adaptation.

In this chapter time resolved microarray data, Northern and Western-blots obtained from rat fibroblast with an inducible, oncogenic Ras are used to identify DUSP6 as a negative regulator that shapes Erk-activation by Ras. We find that Erk-activation is biphasic and, using mathematical modeling, we show that that this is in consistency with the induction of DUSP6. We conclude that in the first few hours after the induction of oncogenic Ras, the dominating phosphatase of Erk is DUSP6. By the induction of DUSP6, cells can show adaptive response to prevent prolonged pro-mitotic signals under physiological conditions. Therefore, DUSP6 can be regarded as a potential tumor suppressor. In the biological system investigated here, the level of Ras-expression is too high, so that DUSP6 cannot terminate the signal at the level of Erk. Nevertheless under parameters corresponding to endogenous expression and more physiological stimuli the system shows significant activation of Erk only for a period of about an hour.

5.1   Introduction

In immortalized rat fibroblasts, expression of oncogenic Ras (H-Ras V12) elicits prolonged activation of Erk and cellular transformation (Zuber et al.,

2000). Oncogenic Ras carries a mutation that inhibits its catalytic activity to hydrolyze GTP. Consequently, Ras is permanently GTP-bound and signals constitutively growth-factor independent to its downstream cascades, including the Raf-Mek-Erk cascade. This way, oncogenic Ras mimics the constant presence of growth factors. Unlike stimulation with growth-factors which induces immediate-early genes, a sustained oncogenic signal yields a deregulation of transcription factors which affect target gene expression in a global way (Sers et al., 2002). Several of the deregulated genes are involved in processes relevant for the development of cancer, such as the cell-cycle. Additionally, some of the induced and repressed genes feed back into the network of Ras-mediated signal transduction and influence its dynamics.

Recently, it became clear that also several dual-specific phosphatases are regulated by Erk-activity, which dephosphorylate Erk and therefore constitute negative feedback loops. For instance, DUSP5 is Erk-regulated (Tchernitsa et al., 2004), and the activation of the Raf/Mek/Erk cascade is sufficient to induce DUSP1 and DUSP4 (Brondello et al., 1997). In Chapter 7 SRF and CREB are discussed as potential transcription factors that control the induction of these phosphatases. In addition, DUSP6 is induced by an yet unknown mechanism (Tullai et al., 2004), possibly also via SRF as unveiled by bioinformatics analysis (Faust, 2005) or involving cross-talk with Akt (Kawakami et al., 2003). Despite of its role in development, the precise physiological role of the dual-specific phosphatases is still unknown, as e.g. DUSP1-knockout mice show no phenotype (Echevarria et al., 2005; Dorfman et al., 1996). Recent findings suggests that induction and stabilization of DUSP1 provides the means for terminating the p38 and Jnk MAPK-cascades, but has no effect on Erk activity (Zhao et al., 2005). These feedback loops may mediate an adaptive response of the cell to prevent prolonged activation of pro-mitotic signals. DUSPs might therefore be tumor suppressors (Warmka et al., 2004; Furukawa et al., 2003). From a theoretical perspective, negative feedback loops have been studied in the MAPK-cascade. Asthagiri and Lauffenburger (2001) showed that inhibition of an adapter protein can yield adaptation. Both Brightman and Fell (2000) and Schoeberl et al. (2002) investigated receptor-mediated feedback loops for signal termination. Bhalla et al. (2002) identified DUSP1 (MKP-1) as a negative regulator that switches the network between a bistable and a mono-stable state. However, the single-cell analysis in the previous chapter showed that Erk-activity is not bistable unlike Bhalla et al. (2002) proposed. Therefore, and since DUSP1 is not specific towards Erk, the role of DUSP1 proposed by Bhalla et al. (2002) is rather unlikely.

In this chapter data from immortalized rat fibroblasts (IR-cells) that are transfected with an inducible oncogenic H-Ras V12 are analyzed. The expression of H-Ras V12 is under the control of an IPTG-sensitive promoter

A   B

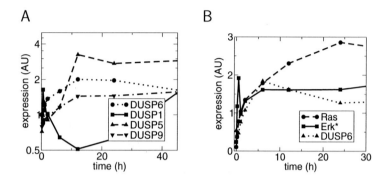

Figure 5.2: A: Time series of the mRNA-expression of several dual-specific phosphatases from Microarray data after induction of oncogenic Ras. B: Time-series of Ras protein expression, Erk activation, and DUSP6 mRNA expression obtained from Western and Northern-blots. Lines are only to connect points of same type.

(Liu et al., 1992). This systems allows time-resolved measurements of the activity of downstream-cascades and of changes in gene expression in the process of cellular transformation. After the induction of Ras, IR-cells overcome density-dependent growth arrest, they acquire a spindle-shaped morphology (see Fig. 5.1) and gain the potential to grow anchorage-independently (Sers et al., 2002). On the genomic level, the expression of about 80 target genes is changed significantly. In the following time series of microarray measurements, Western and Northern-blots after the induction of Ras are utilized to investigate the dynamics of phosphatase expression. A mathematical model is constructed that is consistent with the data and can explain the dynamics of Erk-activation after the induction of H-Ras V12 by induction of DUSP6. The model predicts that the induction of DUSP6 can terminate the signal under physiological conditions.

5.2   The induction of phosphatases

The dephosphorylation of Erk is performed by relatively unspecific constitutively expressed phosphatases such as PP2A and HePTP (Alessi et al., 1995), and by specific and highly regulated dual-specific phosphatases (DUSPs), which can dephosphorylate threonine 183 and tyrosine 185 of Erk. Several

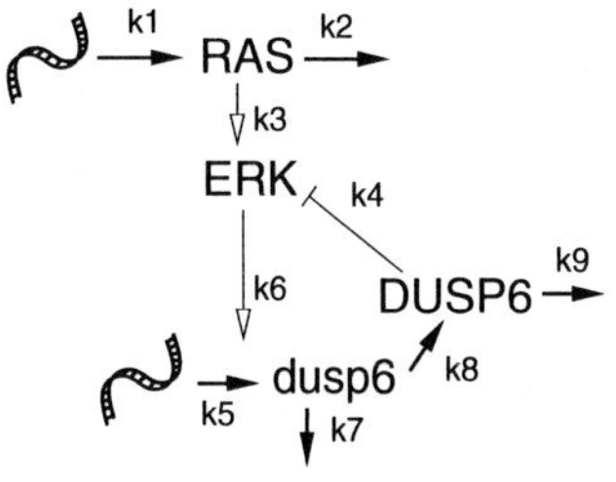

Figure 5.3: Sketch of the model describing Ras induction, Erk activation and DUSP6-regulation. Thin arrows describe regulation, thick arrows show transcription, translation and degradation.

of the DUSPs show a high specificity toward Erk, such as DUSP3, DUSP6, DUSP5 (Farooq and Zhou, 2004). The expression of DUSP1, DUSP5, and DUSP6 was monitored in IR-cells after the induction of oncogenic Ras with micro-arrays and Northern blots (see Fig. 5.2). DUSP6 is immediately upregulated upon induction, followed by an upregulation of DUSP5. DUSP1, in contrast, was first downregulated and was then induced after more than 24 hours. The data suggest that DUSP6 mediates an immediate adaptive response upon Erk activation, and that DUSP5 regulates Erk activity at later time points.

Erk-activation was immediately elevated after induction of Ras and shows a biphasic response with a first peak at 30 min (see Fig. 5.2B). In the following it is investigated whether this biphasic response can be due to the induction of the phosphatase DUSP6.

A first model

To test the idea that DUSP6 causes the biphasic response of Erk after the induction of Ras, a model is constructed (sketched in Fig. 5.3). The expression of oncogenic Ras is assumed to start at time $t = 0$, i.e. at the time the inductor IPTG is given to the medium. Then Ras is expressed with a constant rate and degraded:

$$\frac{\mathrm{dRas}}{dt} = k_1 - k_2 \, \mathrm{Ras} \, . \tag{5.1}$$

Subsequently, Ras promotes the activation by phosphorylation of Erk via Raf and Mek, and Erk is dephosphorylated by DUSP6 (DUSP6$_{\mathrm{prot}}$). Which fraction of Erk is actually activated is not accessible from the western-blot data since the total Erk-concentration is not available. By assuming that Erk is weakly activated, the total concentration as parameter can be eliminated

(Heinrich et al., 2002). Thus, the dynamics of activated Erk is described as follows:

$$\frac{d\text{Erk}^*}{dt} = k_3 \, \text{Ras} - k_4 \, \text{Erk}^* \times \text{DUSP6}_{\text{prot}} \, . \tag{5.2}$$

Finally, activated Erk (Erk^*) enhances the transcription of DUSP6 ($\text{DUSP6}_{\text{mRNA}}$), which is then translated into the protein ($\text{DUSP6}_{\text{prot}}$), both are degraded.

$$\frac{d\text{DUSP6}_{\text{mRNA}}}{dt} = k_5 + k_6 \, \text{Erk}^* - k_7 \, \text{DUSP6}_{\text{mRNA}} \, , \tag{5.3}$$

$$\frac{d\text{DUSP6}_{\text{prot}}}{dt} = k_8 \, \text{DUSP6}_{\text{mRNA}} - k_9 \, \text{DUSP6}_{\text{prot}} \, . \tag{5.4}$$

This model was then fitted to the time-courses of Ras-expression, Erk-activity and the mRNA expression of DUSP6.[2]

To perform a maximum likelihood fit, an error-model needs to be established. However, estimation of the variance in the data is not straight forward for this experiment, as repetitions of this particular experiment showed that the IPTG-sensitive promoter led to different expression rates of Ras and often showed a time-lag before promoting the expression of Ras. Therefore, replicas of the time series could not be used to estimate the variance. In other Western-blot data for this system, repetitions of the experiment had a standard deviation in the order of 20%. In the following an error model with additive and multiplicative Gaussian error components was taken. To estimate the amplitude of these two error components the model was first fitted to the data.[3] Two *ab initio* error-models were taken, one with $\sigma_{add} = 20\%$ additive error of the mean value, the other one with both $\sigma_{add} = 10\%$ additive error and $\sigma_{mul} = 10\%$ multiplicative error. Both error models resulted in residuals with an 10% additive and 10% multiplicative component. Therefore, in the following an error model with both 10% additive and multiplicative noise was used.

To assess whether the parameters can be reasonably estimated, a bootstrapping procedure was applied. First the model was fitted to the data and subsequently one hundred datasets were generated by adding noise to the best fit values. The model was then fitted to the one hundred bootstrapping data sets and the coefficient of variation of the fitted parameters

---

[2]The MATLAB function lsqnonlin was used to perform the least square fit. The ordinary differential equations were integrated using ode45. Initially, all parameters were set to unity. Additionally a least-square fit was performed using the minuit function of cernlib with lsode as integrator, which gave a fit with similar parameters. To test global convergence, random initial conditions for the parameters were used, which yielded no significant reduction of the sum of squared deviations.

[3]The simplified version of the model discussed in the next section was used here.

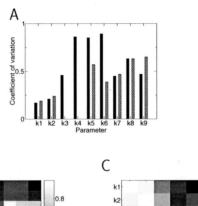

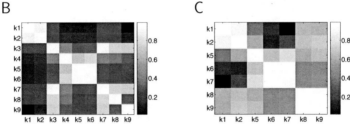

Figure 5.4: A: Coefficient of variation of the best-fit parameters in the model with (solid) and without (striped) explicit Erk-dynamics using hundred boot-strapping data sets. B,C: Absolute value of the correlation coefficients of the estimated parameters from hundred bootstrapping data sets. B: For the model with Erk-dynamics. C: For the reduced model where Erk is assumed to be in quasi-steady-state.

were calculated. Especially the parameters k_4, k_5, k_6 show a high coefficient of variation (cf. Fig. 5.4A). Additionally, the correlation coefficients between the parameters were calculated. Fig. 5.4B shows the absolute values of this correlation-matrix. Not surprisingly, the parameters describing the production and destruction of a species (i.e. expression and degradation, or activation and deactivation) show strong negative correlation, since the steady-state is determined by the ratio of these, and since the system approaches a steady state, the ratio can be determined rather well. Additionally, the parameters describing Erk activation dynamics (k_3) strongly correlate with the parameter k_7, k_8, k_9 that describes the dynamics of DUSP6. This can be explained as follows. First, the time-intervals between the data points are too long to capture the dynamics of Erk, which happens on a time scale about 5 min, and therefore Erk is in quasi-steady-state, and only the ratio $(k_3 \, \text{Ras})/(k_4 \, \text{DUSP6}_{\text{prot}})$ determines the activity of Erk (see below). Second, as there is no antibody available for DUSP6, there is no time series data for the protein. Therefore the level of expressed DUSP6 is arbitrary, and as only the ratio $(k_3 \, \text{Ras})/(k_4 \, \text{DUSP6}_{\text{prot}})$ matters, the expression level $\text{DUSP6}_{\text{prot}}$ cannot be determined from the data.

Reducing the model

As the dynamical details of Erk-activation are not captured by the data, the model is further reduced. First, a quasi-steady-state approximation is used to eliminate activated Erk as a dynamic variable:

$$0 \;=\; k_3 \, \text{Ras} - k_4 \, \text{Erk}^* \, \text{DUSP6}_{\text{prot}} \,, \tag{5.5}$$

$$\rightarrow \quad \text{Erk}^* \;=\; \frac{k_3 \, \text{Ras}}{k_4 \, \text{DUSP6}_{\text{prot}}} \,. \tag{5.6}$$

Second, as the scale of $\text{DUSP6}_{\text{prot}}$ is arbitrary, the factor $\frac{k_3}{k_4}$ can be subsumed in the translation rate of DUSP6 (k_8). A problem might arise by this simplification, since activated Erk approaches infinity when the phosphatase concentration drops to zero. The basal expression rate of DUSP6 (k_5) is kept higher than zero to avoid this problem. The reduced model is then given by:

$$\frac{d\text{Ras}}{dt} \;=\; k_1 - k_2 \, \text{Ras} \,, \tag{5.7}$$

$$\frac{d\text{DUSP6}_{\text{mRNA}}}{dt} \;=\; k_5 + k_6 \, \frac{\text{Ras}}{\text{DUSP6}_{\text{prot}}} - k_7 \, \text{DUSP6}_{\text{mRNA}} \,, \tag{5.8}$$

$$\frac{d\text{DUSP6}_{\text{prot}}}{dt} \;=\; k_8 \, \text{DUSP6}_{\text{mRNA}} - k_9 \, \text{DUSP6}_{\text{prot}} \,. \tag{5.9}$$

Fitting this reduced model to the data yields more robust parameter estimation (compare Fig. 5.4A,C). The dynamics of the system using the estimated parameters is well in agreement with the measured data except that it does not reflect the amplitude of the first peak of Erk-activation, as shown in Fig. 5.5. Additionally, Ras-expression systematically deviates from the model, it seems that the expression rate decreases with time or the degradation rate increases with time. However, another explanation might be that the exposure time for the Western-blot film was such that the film was saturating. As the absolute concentrations are not determined in the Western-blots, the parameters k_1, k_5, k_6 and k_8 can only be given in arbitrary units per hour, these parameters cannot be compared to biochemical

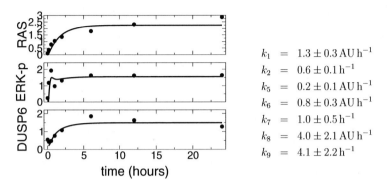

Figure 5.5: The dynamics and fitted parameters of the reduced model. The errors are standard-deviations estimated by bootstrapping

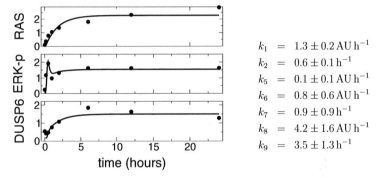

Figure 5.6: The dynamics and parameters of the model with ultrasensitivity.

reaction rates. However, the degradation k_2, k_7 and k_9 rates reflect the degradation rates of Ras, DUSP6 mRNA and protein, respectively. The estimated parameters compare well with literature values: the degradation of DUSP6 mRNA was measured to be $1.3\pm0.5\,\mathrm{h}^{-1}$ (Raghavan et al., 2002). The half-life of the DUSP6 protein was shown to be less than an hour, which corresponds to a rate of more than $0.7\,\mathrm{h}^{-1}$ (Marchetti et al., 2005).

Ultrasensitivity

A possibility for the less pronounced first peak of Erk-activation in the model might be that the signal transduction cascade from Ras to Erk is ultrasensitive. Studies in Chapter 4 revealed that the cascade from receptor to the activation of Erk has an apparent Hill-coefficient of about 4. Since the MAPK-cascade contains two molecules that need to be double-phosphorylated it is possible that this extend of ultrasensitivity is brought about by the MAPK-cascade downstream of Ras (Huang and Ferrell (1996), compare Chapter 2). However, it cannot be judged from this study if it is the cascade downstream of Ras, or both the cascade and the receptor that generate ultrasensitivity in this system. Park et al. (2003) found that activation of PDGF receptor is ultrasensitive with an Hill coefficient of 1.7, such that the Hill-coefficient of 4 found in Chapter 4 can be split into 1.7 for the receptor and about 2 for the cascade, as the Hill coefficient is to some extend multiplicative (compare Chapter 2). Therefore, in the following a moderate ultrasensitivity with a Hill-coefficient of 2 is assumed (although the entire discussion holds also for more pronounced ultrasensitivity). Additionally, the deactivation by DUSP6 is assumed to be ultrasensitive, since dephosphorylation has been reported to occur in a distributed and ordered manner , which can cause ultrasensitivity (Zhao and Zhang, 2001; Ferrell, 1996).

In the following the model is modified such that Erk-activity is given by the ratio of Ras and the DUSP-protein to the power of two, to reflect ultrasensitive activation and deactivation of Erk (compare Chapter 2).

$$\text{Erk}^* = \left(\frac{\text{Ras}}{\text{DUSP6}_{\text{prot}}} \right)^2 . \tag{5.10}$$

Fitting this modified model to the data reproduces the peak of Erk activity during the first two hours more precisely (see Figure 5.6).

5.3   Model selection

Up to now, a model describing the dynamics of Ras, Erk and DUSP6 was constructed. This model was reduced by using a quasi-steady-state approximation and then further refined to include ultrasensitivity of Erk-activation. However, it is not yet addressed whether the reduced model fits the data equally well as the model including the Erk-activation explicitly, and whether the ultrasensitive model fits the data significantly better than the model without ultrasensitivity. To address these two questions, a likelihood-ratio test is applied as it has been proposed by Swameye et al. (2003) for models of signal transduction cascades.

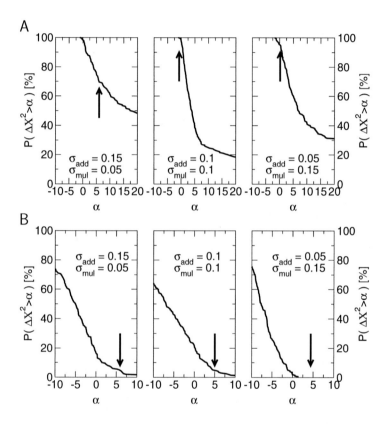

Figure 5.7: The empirical cumulative distributions of χ^2-differences under the null-hypothesis that the reduced model without ultrasensitivity is the correct one. Values larger 0 indecate a higher likelihood of the alternative model. Arrows show the χ^2-difference for the actual data. A: The model containing explicitly the Erk-dynamics as the alternative model. The test rejects this model with $p > 0.6$. B: The model with ultrasensitivity fits significantly better under all error models at $p \leq 0.05$.

The likelihood-ratio test measures the significance in improvement of a fit by a e.g. more complex model over a less complex model. The null hypothesis is that the less complex model is the true model, and that the increased likelihood of a more complex model is only due to over-parameterization, i.e. it is over-fitting the data (Timmer et al., 2004). If the models are nested, the parameters are identifiable, and the parameters are not at the boundary of the parameter space, the log likelihood-ratio follows a χ^2-statistics (consult the detailed discussion in Appendix E). However, as the parameters of the model capturing the Erk-activation explicitly are not identifiable, and as the model with and without ultrasensitivity are not nested (and equally complex), the distribution of the log likelihood-ratio has to be estimated by a bootstrapping procedure, that is outlined in the following (Timmer et al., 2004).

If M_1 is the true model under the null hypothesis, and M_2 is the alternative model, both models are first fitted to the data and the likelihood-ratio is calculated. Subsequently model M_1 is used to generate new datasets by calculating time series and adding noise from an error model. Both models are then fitted to these bootstrapping time series and their likelihood-ratios are computed, which gives an estimate for the likelihood-ratio distribution. If model M_1 is the true model, the true likelihood-ratio should be consistent with this estimated distribution. In consequence, the null-hypothesis is rejected if the likelihood-ratio from the true data-set is not consistent with the hypothesis that it is randomly sampled from the likelihood-ratio distribution estimated bootstrapping procedure at a certain significance level.[4]

A crucial point in this test is the error model that is both used to create the data set from the model and to calculate the likelihood. In this chapter it is assumed that the measurement error follows a Gaussian normal distribution. As stated before, for this experiment the magnitude of the error could not be measured. Therefore, several error models were tried: $\sigma_{add} = 5\%$ and $\sigma_{mul} = 15\%$; $\sigma_{add} = 10\%$ and $\sigma_{mul} = 10\%$; $\sigma_{add} = 15\%$ and $\sigma_{mul} = 5\%$. Here σ_{add} is the standard deviation of the additive noise in percent of the mean value of this variable and σ_{mul} is the standard deviation of the multiplicative noise.

As outlined in Appendix E, the logarithm of the likelihood-ratio corresponds to the differences in χ^2-values of the fits except for a constant, and therefore the χ^2-values can be compared instead of using the likelihood-ratios. Fig. 5.7A shows the empirical distribution of the χ^2-differences for the null

---

[4]The maximum likelihood fit was performed using the minuit library with setting all initial parameters to one. Although this does not guaranty that one finds a global optimum, this should not bias the analysis, as also the observed likelihood ratio was calculated using the same procedure.

hypothesis assuming that the reduced model is the true model. Under all four error models, the true χ^2-difference is not significantly larger. Therefore, the reduction of the model by the quasi-steady-state approximation of the Erk-dynamics is justified. In Fig. 5.7B the empirical distribution of χ^2-differences of the reduced models without and with ultrasensitivity are shown, where the model without ultrasensitivity is the true model under the null hypothesis. Here, the observed difference is significantly larger than if the null hypothesis would be true, therefore the ultrasensitive model fits better.

Another popular criterion to decide which model to take is the Akaike Information Criterion (AIC). AIC provides an estimate of the distance (in this case: the Kullback-Leibler information) between the fitted model and the true model. It is defined as:

$$AIC = -2\mathcal{L}(\theta|D, M) + 2N \,, \qquad (5.11)$$

where $\mathcal{L}(\theta|D)$ is the logarithm of the maximum likelihood, and N is the number of parameters. The term 2N penalizes more parameters assigning the lowest AIC to the most parsimonious model. Similarly as the likelihood-ratio test, AIC needs only the χ^2 values of the best fits to compare models, as in the case of Gaussian error, the AIC can also be expressed as:

$$AIC = \chi^2 + 2N + c \,. \qquad (5.12)$$

Here χ^2 is the minimal χ^2 value of a model, and c is a constant, that is independent of the model. According to Akaike, the most parsimonious model is the model with the lowest AIC. In the case of the models discussed in this section, AIC would also favour the reduced model including ultrasensitive Erk activation, as the ultrasensitive model always has a lower χ^2-value as the equally parameterized reduced model without ultrasensitivity. Additionally, the model with explicit Erk-activation is only favored in the case of the error model with $\sigma_{add} = 0.15$ and $\sigma_{mul} = 0.05$, where the difference of χ^2-values exceeds $2\Delta N = 4$. A disadvantage of AIC is that it is not valid if the parameters are at the boundary of the parameter space or if the parameters are not identifiable. Additionally it favors bigger models (discussed in Timmer et al. (2004)), where AIC would reject the null-hypothesis in 15% of all cases by chance) and is therefore more liberal as the likelihood-ratio test.

5.4 Adaptive response

Negative feedback loops can mediate perfect adaptation through integral feedback control (Yi et al., 2000). In such a system the integrated difference between the desired steady-state output and the actual output is fed

A   B   C

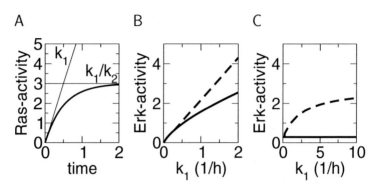

Figure 5.8: A: Parameters k_1 and k_2 describe the activation kinetics of Ras. B: Peak (dashed line) and steady-state (solid line) activity of Ras in the model including ultrasensitivity for the induction of Ras (fixed parameter value for k_2). C: Peak (dashed line) and steady-state (solid line) for different activation kinetics of Ras (the steady-state activation of Ras is kept constant, $k_1/k_2 = 0.25$).

back into the system. Through this, the output shows only transient response towards changes of the input signal, whereas the steady state remains unaffected. Additionally, many systems with negative feedback loops show partial adaptation (Lauffenburger, 2000). After introducing a stimulus and holding it constant, the response of such a system will exhibit an increase to a maximum followed by a decrease to an intermediate level. In the MAPK-cascade, both behaviors have been found (Huang and Ferrell, 1996; Asthagiri et al., 1999b,a), suggesting that different feedback loops act in this pathway resulting in different behavior (Lauffenburger, 2000; Blüthgen and Herzel, 2001).

The negative feedback loop investigated here does not mediate such perfect adaptation, as the steady-state of the output - activated Erk - depends on the input, i.e. the activated Ras (Fig. 5.8B). The peak of activation exceeds the steady-state level only moderately. Nevertheless, the endogenous pool of Ras proteins that is involved in signal transduction under physiological conditions is much smaller than the amount of ectopic Ras expressed in the experiment discussed in this Chapter.[5] In addition, physiological Ras

---

[5]Within 72 hours after IPTG, the amount of Ras rises by a factor of 25. As the antibody detects the normal Ras plus the oncogenic Ras, the ectopic Ras is about twenty-fold higher than the endogenous Ras.

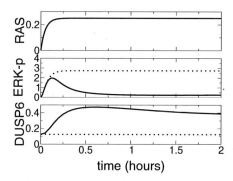

time (hours)

Figure 5.9: Time series of Erk activation with parameters corresponding to physiological Ras-activation ($k_1 = 5\,\text{AU}\,\text{h}^{-1}$ and $k_2 = 20\text{h}^{-1}$) show that Erk shows only transient response with induction of DUSP6 (solid lines), whereas the response of Erk is sustained when DUSP6 is not induced (dotted lines).

activation by growth factors is more rapid. In the following the model is used to investigate whether the induction of DUSP6 can mediate Erk-adaptation to prolonged Ras-activation under more physiological stimuli.

Parameter k_1 describes the initial slope of Ras activation, k_2 determines the time scale of the dynamics, and the ratio of both specifies the steady-state that is approached (compare Fig. 5.8A). These parameters are changed such that activation occurs on a time scale of 3 minutes, and the steady-state activity of Ras is assumed to be about 10-fold smaller when compared to the experiment described above ($k_1 = 5\,\text{AU}\,\text{h}^{-1}$ and $k_2 = 20\,\text{h}^{-1}$). To simulate different activation kinetics of Ras, the parameters k_1 and k_2 are simultaneously altered such that their ratio remains constant. As expected, steady-state Erk-activity is unaffected by these changes, but the transient peak of Erk-activation increases with accelerated Ras-activation, and can exceed the steady state significantly (see Fig. 5.8C).

Time-series of this system illustrate that the induction of DUSP6 can protect the cells from prolonged Erk-activation (see Fig. 5.9). Upon activation of Ras, the system shows significant Erk-activation for about 20 minutes after which Erk is nearly deactivated. The maximal activity of Erk would be similar without DUSP6 induction. However, Erk activation would be prolonged (compare dotted lines in Fig. 5.9).

Fig. 5.10 shows that both under physiological conditions and over-expression of oncogenic Ras, the induction of DUSP6 controls both the timing

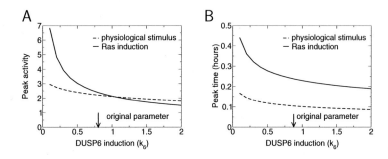

Figure 5.10: Both maximum peak activation (A) and time of peak activation (B) show a similar dependence on induction of DUSP6.

and the amplitude of peak Erk-activation. Thus, phosphatase induction as a terminator of continuous signals does control both the strength and length of the signal transduced. Previous studies by Hornberg et al. (2005b) and Heinrich et al. (2002) showed that constitutively expressed phosphatases mainly affect signal duration when signals are transient. Here, the regulation of a phosphatase effects both duration and peak-activity, especially for strong stimulation.

5.5 Discussion

Research on signal transduction was mainly focused on how a signal is turned on, while the mechanisms by which it is terminated are less well investigated (Marshall, 1995). However, in recent years the termination of Ras-mediated signaling has been addressed by several theoreticians. Signal termination has been thought to be due to desensitization and internalization of the receptor, and modification of adaptor proteins (Schoeberl et al., 2002; Asthagiri and Lauffenburger, 2001), and by phosphorylating Raf at a second residue (Asthagiri and Lauffenburger, 2001; Blüthgen and Herzel, 2001; Bhalla and Iyengar, 1999). However, it becomes more and more clear that phosphatases have an important influence on the dynamical features of kinase-cascades (Heinrich et al., 2002; Mayawala et al., 2004; Hornberg et al., 2005b). Additionally, it becomes apparent that the specific phosphatases for Erk are highly regulated on the transcriptional level (Tchernitsa et al., 2004; Brondello et al., 1997; Tullai et al., 2004). Also the post-translational regulation of and by dual specific phosphatases has been addressed, and time constants have been measured (Brondello et al., 1997; Marchetti et al., 2005; Pouyssegur et al., 2002).

Here, time series data after the induction of oncogenic Ras has been used to get insight into the regulation of Erk by phosphatases, as this experimental system does not require receptor-activation. This way, the feedbacks that target the receptor or adapter molecules could be neglected, and the influence of the induction of DUSP6 on the dynamics of Erk could be investigated. A mathematical model was constructed that could explain the dynamics of Erk-activation by the induction of DUSP6 after the induction of oncogenic Ras.

If the stimulus was chosen to be in a more physiological range, e.g. with fast activation and lower absolute concentration of Ras, the negative feedback loop via DUSP6 terminates Erk-activation. Therefore, DUSP6 might act as a signal terminator for growth-factor independent signals such as calcium or phorbol 12-myristate 13-acetate (PMA) induced signaling. It may furthermore function as a backup providing redundancy if receptor desensitization in growth-factor induced signal transduction fail. Such redundancy in signal termination may be of importance in the control of an cancerogenic kinase Erk. Additionally it might function as a protector against signals received from endogenous oncogenic Ras.

This analysis suggests the following experiments to confirm the predicted role of DUSP6. First, knock-down of DUSP6 with siRNA upon induction of Ras should give elevated levels of activated Erk and a monophasic response. Second, under physiological growth-factor independent stimuli such as activation of PKC by PMA, knock-down of DUSP6 should result in prolonged Erk-activation. Additionally, pre-treatment with cyclohexamide should give rise to prolonged Erk-activation upon physiological stimuli.

Other phosphatases despite DUSP6 are also regulated. The transcriptional repression of DUSP1, which is not specific for Erk, might open the door for other pathways. As DUSP1 preferentially inactivates p38 MAPK and Jnk, its downregulation sensitizes these pathways which may lead to stress-responsive MAPK-mediated apoptosis. The transcriptional regulation of DUSPs might be a major means of rearranging the MAPK signal transduction network.

6

Functional profiling of Ras-responsive genes

The results of this chapter have been partially published in Blüthgen et al. (2005a) and to be published in Jürchott et al. (2005). Further application of the framework are discussed in Bandapalli et al. (2005). The GOSSIP software has been developed in cooperation with Dieter Beule. A review of this method is to be published in Blüthgen et al. (2006a). Microarray data sets from Tullai et al. (2004) and unpublished microarray data from the Schäfer/Sers lab are used.

Synopsis

Increasingly used high throughput experimental techniques, like DNA or protein microarrays, give as result groups of interesting, e.g. differentially regulated genes. Groups of Ras-responsive genes have been determined in high throughput experiments for both growth-factor induced signal transduction and for signalling due to oncogenic Ras. Using inhibitors, these groups have been further refined by analyzing the specific pathway they respond upon. In this chapter we address the question whether these gene groups correspond to specific biological processes. With the systematic functional annotation provided by the Gene Ontology the tool is now in hand to automatically address this question without prior hypothesis. Several approaches exist to annotate and help with the interpretation. Nevertheless, the definition of statistical significance for a certain e.g cellular process in the annotation of a group of genes is still an open question. In answering this question, multiple testing issues must be taken into account to avoid misleading results.

In this chapter a statistical framework is described that tests whether functions, processes or locations described in the Gene Ontology are significantly enriched within a group of interesting genes when compared to a reference gene group. First we define an exact analytical expression for the expected number of false positives that allows us to calculate adjusted p-values for the control of the false discovery rate. Then the robustness of the framework with respect to the exact gene group composition is analyzed. Finally, gene groups that are regulated by Ras mediated signal transduction are analyzed. Here, gene groups from microarray experiments with growth-factor stimulation, induction of oncogenic Ras and inhibition of specific pathways are analyzed. Interestingly, processes related to the immune system seem to be differentially regulated upon growth-factor stimulation and induction of oncogenic Ras, where genes from the major histo-compatibility complex (MHC-I) are down-regulated.

6.1   Introduction

With the advent of genome-wide screening experiments, like microarray studies (Duggan et al., 1999), and 2D protein gel analysis (Klose et al., 2002), researchers frequently face the task of interpreting the biological function and relevance of gene groups, e.g. groups of differentially expressed genes. Even if the individual genes are annotated this interpretation task remains laborious and complex. Analysis of gene groups using an ontology is a promising starting point for an automated biological profiling beyond the single-gene level. An ontology specifies a controlled vocabulary and the relations between the terms within the vocabulary. This concept is widely used to systematically represent knowledge for further analysis. For molecular biology the Gene Ontology Consortium provides the Gene Ontology (GO) as an international standard to annotate genes and gene products (Ashburner et al., 2000). The Gene Ontology defines more then 16500 terms to describe the molecular functions, affiliation with processes and locations of a gene. These terms are arranged in a hierarchy such that a term that is deeper in the hierarchy is more specific and implies more general terms upward in the hierarchy (compare Fig. 6.1).

In this chapter, a statistical framework is described, that utilizes GO to test whether a molecular function, biological process, or cellular location (which we call a "term" in the following) is significantly associated with a group of interesting genes. The definition of statistical significance is a major challenge due to the large number of terms which need to be tested. The use of single test p-values is only justified if we test whether a single term is associated with a specific gene group. However, in genome-wide screening experiments the situation is fundamentally different: the current GO includes over 16500 terms, out of which typically several thousand terms appear in the annotation of an investigated gene group and have to be tested. If one performs that many tests, problems arising from multiple testing cannot be left aside. Namely, even when we apply a very conservative threshold like $p < 0.001$ a few terms will be reported to be associated with the test group by sheer chance. This phenomenon is known as false positives or type-I-error. One solution for this problem is to calculate adjusted p-values. These adjusted p-values control the number of false discoveries in the entire list and can be used similarly to normal p-values for single tests (Dudoit et al., 2003). There are several standard methods to calculate adjusted p-values, like re-sampling and multi-step estimations. However, for functional profiling of gene groups the adjusted p-values obtained with standard multiple testing correction methods are unsatisfactory because they are either not precise enough, or they require major computational efforts.

This chapter is structured as follows. First, a novel statistical framework is described that is based on an analytical estimate of false positives, and this framework is compared to earlier approaches[1]. Subsequently, we investigate the robustness of our framework with respect to the exact gene group composition. This is of special importance as in the next chapter the framework is used to profile very "noisy" gene groups. Then the framework is used to investigate the biological processes regulated by growth-factor induced signal-transduction on the one hand and by dis-regulation due to oncogenic Ras on the other hand.

6.2 Statistical framework

Data preparation

In order to profile gene groups four data sources are required: a test group of genes (e.g. up-regulated genes), a reference group (e.g. all significantly expressed genes), GO annotations for these genes, and the Gene Ontology graph. Many chip-manufacturers provide GO annotations for the genes covered by their chips[2]. If they are not available, gene groups can be annotated using tools like HomGL (Blüthgen et al., 2004). The current version of the Gene Ontology can be downloaded from the website of the Gene Ontology

---

[1]This framework has been implemented in a software-package GOSSIP, available at http://gossip.gene-groups.net/

[2]e.g. Affymetrix (TM), see http://www.affymetrix.com/support/

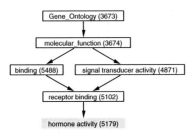

Figure 6.1: Part of the directed acyclic graph (DAG) representing the Gene Ontology. Annotations are usually given as terms close to the leafs of the DAG, e.g. term 5179 hormone activity, implying a series of more general terms (identifiers 5102, 5488, 4871, 3674, 3673).

Consortium. The Gene Ontology can be represented as a directed acyclic graph (DAG) where the nodes represent the terms (Bard and Rhee, 2004). Annotations are usually given as terms within the DAG implying a series of more general annotations upward in the GO graph, as illustrated in Figure 6.1.

Statistical framework

For each term in the ontology we ask whether this particular term is enriched in the test group as compared to the reference group. To test this we categorize each gene in two ways: first, whether it is annotated with the term under consideration or not, and second, whether it belongs to the test group or not. Based on these categories we build a 2×2 contingency table of gene frequencies for each term. Figure 6.2 shows the structure of such a contingency table. Using Fisher's exact test[3] we compute p-values that allow to detect and quantify associations between the two categorizations. Fisher's exact test is based on the hyper-geometric distribution, and works in a similar way as the χ^2-test for independence. The χ^2-test provides only an estimate of the true probability values, and it is not accurate if the marginal distribution is very unbalanced or if we expect small frequencies (less than five) in one of the cells of the contingency table. Both situations are typical for the task and data under consideration. Although Fisher's exact test can in principle quantify the reduction of a term with respect to the reference group we focus on enrichment or association. A reduction is unlikely to be detected in typical data sets, since the test group is usually much smaller than the reference group.

To control the number of false discoveries, we determine adjusted p-values corresponding to the false discovery rate (FDR) that quantifies the expected

---

[3]http://home.clara.net/sisa/fishrhlp.htm

Figure 6.2: 2×2 contingency table of gene frequencies that is calculated for each term. Each gene in categorized in two ways: whether it belongs to the test group and whether it is annotated with the term under consideration. This figure shows the contingency table for the term "mitotic cell cycle" in the group of genes that are up-regulated during the G2-phase of the cell cycle. In total, 14480 genes are in the reference group. 437 genes are annotated with this specific term, 20 of them are in the G2-phase gene group. 14043 are not annotated with this term. Out of this, 65 are in the test group. The number of genes in the test group is 85.

		Test Group Yes	No	Reference Group
Annotated	Yes	20	417	Zi=437
	No	65	13978	
		T=85		N=14480

portion of false discoveries within the positively tested. If there is no prior expectation about an association between the gene list and any biological process, one might favor the family-wise error rate (FWER, see Appendix D). However, the typical case in profiling gene lists is that one expects some terms to be enriched. In this case the FDR is an adequate measure of false discoveries. Both rates can be reliably estimated by re-sampling simulations, but this method suffers from very long runtime even on modern computers. Alternatively, several approaches exist to estimate the FDR from the single-test p-values (e.g. Benjamini-Hochberg (Benjamini and Hochberg, 1995), and Benjamini-Yekutieli (Benjamini and Yekutieli, 2001), for application of these methods to microarray data, see Dudoit et al. (2003)). These methods are designed to cope with general problems. In the following it is shown that for the specific problem of profiling gene groups, the expected number of false discoveries for a given p-value threshold can be determined exactly by an analytical expression. Consequently, the FDR can be calculated exactly.

Terminology

The number of genes in the reference group is denoted by N, the test group is a subgroup of the reference group, with T being the number of genes within this group. With K we denote the number of GO-terms that annotate genes in the reference group. We index these GO-terms with $i = 1...K$, and Z_i denotes the number of genes in the reference group being annotated by the

GO-term i. In the example of Figure 2 these numbers would be: $N = 14480$, $T = 85$, and $Z_i = 437$.

Analytical estimate of false discoveries

For a given p-value threshold α we obtain the expected number of false discoveries ($\mathrm{NFD}(\alpha)$) by summing over all possible tests with weights according to the probability that they are positive by chance:

$$\langle \mathrm{NFD}(\alpha) \rangle = \sum_{i=1}^{K} \mathrm{Pr}(p_i \leq \alpha). \tag{6.1}$$

Here $\mathrm{Pr}(p_i \leq \alpha)$ denotes the probability that the unadjusted p-value of term i with its marginal distributions matches the threshold α. We use the hyper-geometric distribution $h(j, T, N, Z_i)$ to describe the probabilities of observing j annotations given the marginal distribution (T, N, Z_i). Then $\mathrm{Pr}(p_i \leq \alpha)$ can be calculated by

$$\mathrm{Pr}(p_i \leq \alpha) = \sum_{j}^{p_f(j,T,N,Z_i) \leq \alpha} h(j, T, N, Z_i). \tag{6.2}$$

The hyper-geometric distribution is given by

$$h(j, T, N, Z_i) = \frac{Z_i! T! (N - Z_i)! (N - T)!}{N! j! (Z_i - j)! (T - j)! (N - Z_i - T + j)!}, \tag{6.3}$$

and $p_f(j, T, N, Z_i)$ denotes the p-value of the one-sided Fisher test for j or more annotations in the test group and can be calculated by summing over the hyper-geometric distribution:

$$p_f(j, T, N, Z_i) = \sum_{k=j}^{\min(Z_j, T)} h(j, T, N, Z_i). \tag{6.4}$$

In order to validate our analytical result (Eqn. 6.1) we estimate the number of false discoveries using re-sampling simulations. We keep the reference group fixed with N genes and then select random test groups of size T (compare Figure 6.2). The expected number of false discoveries for a specific p-value threshold α is estimated by the mean number of positive tests in the re-sampling runs. Figure 6.3 shows that analytical and numerical results are in excellent agreement for different test group sizes. The sudden jumps in the number of false discoveries can be traced back to the discrete nature

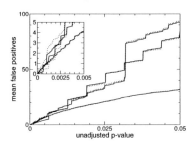

Figure 6.3: Comparison of the analytical results with re-sampling simulations. Expected number of false discoveries calculated in this work (solid line) and estimated by re-sampling runs (dotted lines) of 100 random test sets containing 10, 100 and 500 genes taken from all annotated genes of the Affymetrix HG-133A chip as a reference set.

of the contingency table and its marginal distributions. Note that the correlations between terms induced by the structure of the graph and by the annotation do not influence the mean number of false discoveries but only skew the distribution.

False discovery rate (FDR)

If one expects some terms to be enriched in the test group, controlling the false discovery rate (FDR) is the appropriate method (Dudoit et al., 2003). The FDR gives an estimate of the proportion of the expected number of false discoveries $\mathrm{NFD}(\alpha)$ among all positives $\mathrm{R}(\alpha)$ for a given p-value threshold α:

$$\mathrm{FDR}(\alpha) = \frac{\langle \mathrm{NFD}(\alpha) \rangle}{\mathrm{R}(\alpha)}. \qquad (6.5)$$

Since we can calculate $NFD(\alpha)$, exact determination of $FDR(\alpha)$ is possible, and an adjusted p-value of the portion of false discoveries is given by $p_{FDR}(p) = \min(\mathrm{FDR}(p), 1)$. Each list of terms that fulfills the criterion $p_{FDR}(p) \leq 0.05$ is expected to contain 5% terms that are false discoveries.

Figure 6.4 shows comparisons of the FDR calculated by our methods with other approaches for data sets described in Blüthgen et al. (2005c). Our method is in excellent agreement with the adjusted p-values calculated by re-sampling simulations which provide a reliable estimate of the true FDR. A particularly interesting property of this curve is the sudden jumps of the FDR. These jumps are caused by the discrete values of the p-values in the Fisher's exact test due to the discrete nature of the contingency table. In the example discussed here, some GO-terms annotate only two genes, and their lowest possible p-value is 0.033. Therefore these GO-terms can never be significantly enriched, when a single-test p-value threshold is 0.03. In contrast

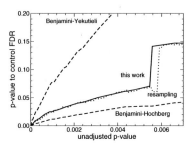

Figure 6.4: The adjusted p-value for the group of genes expressed in the G2-phase of the cell cycle in HeLa cells to control the FDR estimated by our approach (solid) are in excellent agreement with the re-sampling simulations (dotted). Adjusted p-values (dashed) with multiple step Benjamini-Hochberg and Benjamini-Yekutieli are shown with dashed lines.

to our method and the re-sampling simulations, the methods according to Benjamini-Yekutieli (BY) and Benjamini-Hochberg (BH) cannot account for these discrete p-values. Therefore, they cannot reproduce the sudden jumps and do not estimate the FDR precisely for our specific problem. Some authors, like the authors of FatiGO (Al-Shahrour et al., 2004), try to avoid this problem by only testing at a certain level, or like in GeneMerge (Castillo-Davis and Hartl, 2003), limit the search to terms that annotate two or more genes. Both approaches have disadvantages: the first approach neglects that the Gene Ontology is a directed, acyclic graph and the concept of levels can only apply to trees. Additionally, it is unclear whether all terms at a certain "level" are similarly specific. The limitation used in the latter approach is rather arbitrary, as it is not a priori clear, which number of genes need to be annotated by a term that it can get significant. The re-sampling simulations in Figure 6.4 show that the appropriate threshold allowing 5% false discoveries is in this case around p=0.003. The upper limit for the FDR provided by BY is far to conservative: if one would allow 5% of false discoveries in Figure 6.4, the single-test p-value threshold estimated by BY would be around p=0.0005. Thus this method misses significant results. The estimate provided by BH (p=0.0075) is not reliable, yielding more false discoveries then specified by the threshold.

Robustness of biological profiles

When extracting a gene group from high-throughput experiments, one always has to deal with the trade-off between specificity and sensitivity. By increasing the group size, an increased portion of genes is included in the group just by chance and not due to biological or functional reasons. Therefore it is often not clear which genes to include in or exclude from a certain

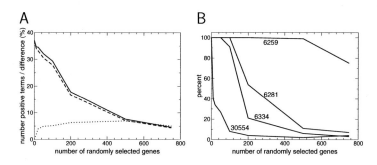

Figure 6.5: The effect of adding randomly selected genes to the test group: to the gene group of cell-cycle regulated genes expressed in the S-phase (221 genes), we add randomly 10, 20, 50, 100, 200, 500 and 750 genes from the reference group and profile these gene groups. (A) The solid line shows the mean number of significant terms in 100 runs. The dashed line displays the number of those terms, which are significant in the initial group. The dotted line shows the difference between both lines in percent. (B) Percentage of runs where the terms with identifier 30554, 6334, 6281 and 6259 are still detected as significant.

gene group, e.g. by choosing a specific threshold. In principle these falsely assigned genes might result in falsely discovered GO terms.

In the following we show that the results of our method do not depend critically on the precise composition of the test group, and are robust with respect to a high portion of genes which are accidentally assigned to the group. Here we assume that the falsely assigned genes are randomly distributed on the microarray. This assumption does not necessarily hold for gene lists obtained with rather naive microarray analysis, since e.g. the absolute signal or cross hybridization might induce a bias. However, there are approaches to limit such a bias, for example by estimating the variance for each gene from replications (Dudoit et al., 2002), or by variance stabilization (Huber et al., 2002). We address the robustness by profiling the gene group of 221 cell-cycle regulated genes expressed in the S-phase (Whitfield et al., 2002) and adding randomly selected genes to this group. Initially, 37 terms are reported to be significant (FDR<0.05). Subsequently 10, 20, 50, 100, 200, 500, and 750 randomly selected genes from the reference group (the microarray) are added to the initial test group and the resulting groups are profiled. We repeated this procedure 100 times and determined how much of the initial profile is preserved. Figure 6.5A shows that the number of significantly enriched terms decreases if we add more and more random genes. However, the profile is remarkable robust: even if we add 200 randomly selected genes, we can still detect 17 of the initially 37 significant terms. Interestingly, from the difference between the solid and dashed line we see that only about 5% of the terms in the resulting profiles are not contained in the initial profile for the S-phase, regardless of the number of randomly selected genes added. These are potentially falsely discovered terms confirming our criterion of the adjusted p-value to control the FDR<0.05. Figure 6.5B shows the robustness of terms with different initial FDR. As expected, terms with a FDR just below the threshold of 0.05, like *adenyl nucleotide binding* (id 30554, FDR=0.044), are unlikely to be detected after adding many randomly selected genes. However, terms with intermediate significance, like *nucleosome assembly* (id 6334, FDR=0.0091), and *DNA repair* (id 6281, FDR=0.00099), can still be detected after adding 100 randomly selected genes. Remarkably, highly significant terms like *DNA metabolism*, (id 6259, FDR=$5.2\,10^{-8}$), are found in 99% of all cases even after adding 500 randomly selected genes. These results show that our framework performs reasonably well even if many randomly selected genes are in the gene groups. Especially, intermediate and highly significant terms will persist. Furthermore, the framework controls the number of falsely discovered terms reliably. We have additionally profiled 12 groups of up-regulated genes of six replications of a microarray experiment, where for each experiment two p-value thresholds for significant up-regulation were used. We found that all profiles include the same terms, and only one out of 12 gene groups resulted into two potentially falsely reported terms among the 15 significant terms. The profile of another group showed an additional, more specific term[4].

6.3   Functional profiles of Ras-targets

In the following, the framework is applied to groups of Ras target genes found by microarray experiment. First, a group of genes responding to growth-factor stimulation is analyzed. This group is further divided into Erk-dependent and Akt-dependent activation (Tullai et al., 2004). Second, gene groups that respond upon induction of ectopic oncogenic K-Ras are profiled. Third, genes that respond consistently upon Mek-inhibition in three colon-carcinoma cell lines expressing endogenous oncogenic Ras or Raf are analyzed (Jürchott et al., 2005).

---

[4]This data is described in Chapter 7.

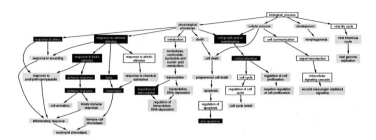

Figure 6.6: Biological processes associated with PDGF-induced genes. Black and gray nodes represent terms associated with the gene group with FDR<0.01 and FDR<0.05, respectively. White nodes display the context of the Gene Ontology.

A   B

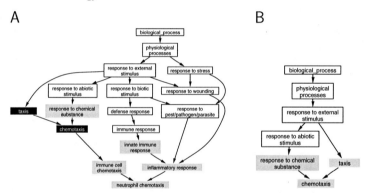

Figure 6.7: Biological processes associated to pathway-specific PDGF-induced genes. A: Processes associated with Erk-activity, B: processes associated with PI3K-activity. Notation as in Fig. 6.6.

Growth-factor induced processes

Platelet-derived growth factor (PDGF) was one of the first mitogens to be identified (Alberts et al., 2002). When blood clots, as it is the case in wounds, platelets incorporated in the clot are triggered to release the contents of their secretory vesicles, which includes PDGF. Cells of diverse types start to proliferate and to close the wound by migration after receiving PDGF. Typically, the presence of a growth-factor rapidly induces about hundred so-called immediate early genes. These genes enable the cells to close the wound, e.g. by entering the G1-phase of the mitotic cell cycle, and inducing cell migration. Also, transcription factors are induced that regulate processes being important in the later phases of wound healing, like entry into the S-phase of the cell cycle. PDGF stimulates a variety of cell types, including fibroblasts, smooth muscle cells, and neuroglial cells (Alberts et al., 2002).

Tullai et al. (2004) investigated growth-factor induced genes in human glioblastoma cells (T98G cells). They stimulated these cells with platelet-derived growth factor (PDGF) and analyzed the changes in gene-expression by DNA microarrays. They found that 74 genes are induced more then 2-fold after 30 min of stimulation with PDGF. Reanalysis of the data showed that only 59 of these genes showed a significant change when a t-test with threshold $p = 0.01$ is applied[5]. The processes significantly associated with these genes are shown in Figure 6.6.[6] Most significantly affected processes are related to *wound healing*, like *cell growth, anti-apoptosis, chemotaxis, immune-* and *stress-response*. Also *regulation of transcription* is significantly enriched in this gene group, highlighting the regulation of late responses by induced transcription factors. Some other processes like *viral infectious cycle* and *morphogenisis* are less obvious. These processes might be false positives, or they annotate genes that are involved in several processes.

Furthermore, Tullai et al. (2004) defined groups of genes that show different behavior when inhibitors of Mek- and PI3K-signaling were applied. They defined genes to be Mek-, and PI3K-dependent if their induction was reduced by more then 50%. Also here, Tullai et al. (2004) did not test whether this reduced induction is significant. Using Student's t-test, a reanalysis of their data yielded 17 genes that required Erk-signaling, 19 genes that required PI3K signal transduction and 40 genes that were significantly effected when both pathways were blocked.[7] Interestingly, *MAP kinase phosphatase activity* is the only term significantly associated with the group of genes that changed when both inhibitors were applied. It seems that the cell responds to both prolonged Erk- and AKT stimulation by up-regulating dual specific phosphatases (DUSPs). This underlines the importance of signal-termination due to DUSPs discussed in the previous chapter. The transcription factors that might regulate this response are investigated in the next chapter. Fur-

---

[5]The raw data were downloaded from GEO (accession numbers GSM18502-GSM18512 and analyzed using Student's t-test. Only these genes were analyzed that had sufficient replicas to estimate the variance, i.e. genes that showed spots well above the background in more then 3 replicas in the experiments with inhibitors, or more then 1 replica in the experiment without inhibitors.

[6]All genes that had sufficient replicas were used as background set.

[7]A threshold of p<0.01 was applied.

thermore, the inhibitors, when applied separately, both showed significant influence on the induction of genes involved in chemotaxis (see Fig. 6.7). Additionally, the Mek-inhibitor significantly changed inflammatory response. Tullai et al. (2004) did only use 2 replicas for each inhibitor, causing a the t-test to be less sensitive, and it is likely that therefore other processes could not be detected as significantly inhibited.

Oncogenic Ras in HEK cells

Oncogenic Ras mimics continuously presentation of growth and survival factors to the receptors. However, the genes induced by oncogenic Ras might only partially overlap with growth-factor induced genes, as late-responses via the induction of transcription factors might dominate under prolonged stimulation with oncogenic Ras, and growth factors might regulate processes independently of Ras. In the following, microarray data from HEK cells with inducible oncogenic Ras will be analyzed for significantly regulated processes. The analysis yielded 251 up-regulated and 274 down-regulated genes due to oncogenic RAS, where 113 and 72 gene were dependent on Erk-activity, respectively.[8] The processes associated with the down-regulated genes are related to *immune response*, and to *extracellular matrix* (see Fig. 6.8). Also genes involved in *phosphate transport* are down-regulated. Processes like *blood coagulation* and *wound healing* are up-regulated as well as *regulation of cellular process*, which confirms the idea that Ras induces several other regulators via transcription. Interestingly, no significantly associated processes with genes up-regulated by Erk are found. It seems that the induction by Erk is less specific under the stimulation of endogenous oncogenic Ras. The genes repressed by Erk were significantly associated with processes related to *extracellular matrix, cell adhesion, phosphate transport* and *muscle development* (compare Fig. 6.9). The appearance of muscle-related genes in the down-regulated set in is surprising for HEK-cells; however, there are many genes related to the extracellular matrix that play a role also in muscles.

A

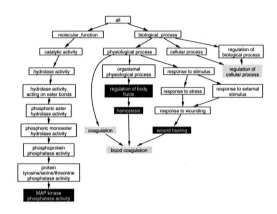

B

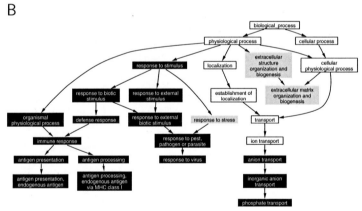

Figure 6.8: Significantly associated processes with the genes up- and down-regulated after the induction of ectopic oncogenic Ras. Notation as before.

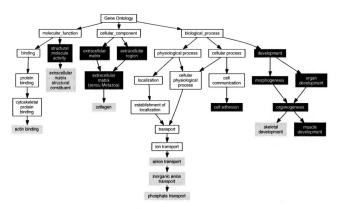

Figure 6.9: Terms significantly associated with genes whose repression by ectopic oncogenic Ras requires Erk-activation. Notation as before.

Oncogenic Ras in tumor cell lines

Jürchott et al. (2005) investigated the transcriptional response of several tumor cell lines upon inhibition of Mek/Erk. They used three colon-carcinoma cell lines, two carrying a mutation in K-Ras (SW480, HCT116) and one with mutated b-Raf (HT29). The cells were grown under normal serum conditions and then the Mek inhibitors U0126 and PD098059 were applied. The idea behind this study was to exclude cell-line-specific effects and to find pathway-specific genes that are dis-regulated by endogenic oncogenic molecules rather than ectopically expressed mutated molecules. As discussed in the previous chapter, ectopic Ras might circumvent tumor suppressors by a higher expression level. Therefore, ectopic and endogenous Ras might have different effects. Jürchott et al. (2005) used Affymetrix HG-U133A microarrays and defined 35 clusters using k-means clustering. They found four clusters (clusters AM02, AM10, AM13, AM23) that showed coherent down-regulation of genes among all cell lines, when the inhibitors were applied.

Figure 6.10 shows the biological processes that are significantly associated with at least one of the four clusters. Remarkably, most processes found are

---

[8]Seven Affymetrix chips were analyzed (3 with induced Ras, 2 with additional MEK-inhibitor U0126, and 2 without induced Ras). The analysis was performed using RMA and LME bioconductor and R (Gentleman et al., 2004). An FDR-threshold of 0.01 was chosen and calculated with Benjamini-Hochberg (Benjamini and Hochberg, 1995). Erk-dependent genes are genes that were coherently regulated in Ras-induced cells when compared to cells without induction and cells with U0126 inhibitor.

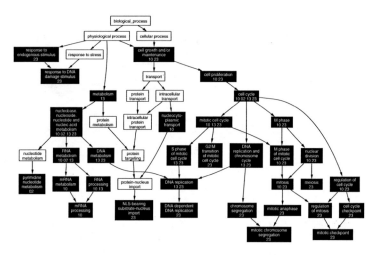

Figure 6.10: Black nodes represent processes associated with genes that are in a cluster that consistently dependents on Erk-activity in three colon-carcinoma cell lines (FDR<0.05) (Jürchott et al., 2005).

related to the cell-cycle.

6.4   Discussion

Statistical Framework

In this chapter a statistical framework is proposed to find molecular functions, biological processes and cellular locations significantly associated with gene groups. This approach allows an unbiased biological profiling of gene groups beyond the single gene level. Special attention is payed to the multiple testing problem, and an exact determination of the expected number of false discoveries is given. The framework is validated with re-sampling simulations. Adjusted p-values can differ from the single test p-values by a factor of more than 10^4. In contrast to re-sampling simulations, which are very slow (up to hours on a typical workstation), the approach presented in this chapter needs just a few seconds. Comparison with other approaches shows that the method of Benjamini-Yekutieli is too conservative, resulting in adjusted p-values which are 2-6 times higher. Thus, this methods would miss many significant results. The Benjamini-Hochberg estimate is not reliable for the problem under consideration.

There are several software implementations available to profile gene groups using Gene Ontology, including Onto-Express (Draghici et al., 2003), EASE+David (Hosack et al., 2003; Dennis et al., 2003), GoSurfer (Zhong et al., 2003), GoMiner (Zeeberg et al., 2003), GeneMerge (Castillo-Davis and Hartl, 2003), FatiGO (Al-Shahrour et al., 2004), and GOstat (Beissbarth and Speed, 2004). The software packages listed before, with the exception of Onto-Express (free version) and GoMiner, have some multiple testing correction. We conclude from our studies, that results of applications that do not use multiple testing corrections (Onto-Express and GoMiner) are hard to be interpreted since false positive predictions dominate. On the other hand, the packages using appropriate standard multiple testing corrections (Bonferroni in Gene Merge, Benjamini-Yekutieli in FatiGo, GoStat, GOSurfer) do give control of the number of false discoveries, but are too conservative and therefore have less power. Only EASE+David uses a jackknife procedure similar to re-sampling to correct for multiple testing, which can give more robust scores although they cannot be interpreted as adjusted p-values.

Furthermore, we show that our method is robust with respect to randomly assigned genes in the gene groups. Therefore the biological profiles do not depend critically on the details of the prior analysis, e.g. threshold, parameters of cluster analysis methods, and normalizations.

Ras regulated processes

Ras has a central role in growth-factor induced signal transduction. This chapter discusses functional profiles for genome-wide microarray data from: growth-factor stimulation, cells stably transfected with oncogenic Ras and endogenous oncogenic Ras. These functional profiles show that Ras-mediated signal transduction can have opposing effects. While growth factors stimulate genes involved in immune-response, this process is down-regulated when endogenous oncogenic Ras is induced. The profiles of genes that depend on Erk in the three colon-carcinoma cell lines are remarkably homogeneous and contain only terms related to cell-cycle, highlighting the importance of pro-proliferating effects of endogenous oncogenic Ras. Surely, the differences between the regulated processes might result from the different cell types. However, it might also be that these differences are due to the different signals. While growth-factors signals act only transiently and might stimulate pathways independently of Ras, oncogenic Ras signals prolonged and might signal to only a part of the cascades that are stimulated by growth-factors. Additionally, the difference in signal strength between endogenously and ectopically expressed Ras might cause differences in the response as ectopic Ras might bypass negative regulators (compare previous chapter).

It is also notable that both growth-factor stimulation and induction of oncogenic Ras induce MAP kinase phosphatases by growth-factor signaling via both the Raf-Mek-Erk pathway and the PI3K-Akt pathway. The transcription factors that might account for this regulation are discussed in the next chapter, where the framework developed in this chapter is applied to predict the functional targets of transcription factors by profiling gene lists that display clusters of binding sites in their upstream regions (Kielbasa et al., 2004b; Blüthgen et al., 2005c). Although the gene groups that display clusters of binding sites are dominated by non-functional sites, the robustness of the statistics discussed in this chapter allows to make reliable predictions for the true functional targets.

7

Predicting Ras- and Erk-regulated processes

The method described in this chapter is joined work with Szymon Kielbasa. The results for the LPS-induced genes and further analysis have been published in Kielbasa et al. (2004b) and Blüthgen et al. (2005c).

Synopsis

The major target of Erk is the activation of several transcription factors. Thereby Erk controls the expression of a few hundred so called immediate-early genes. In the previous chapter microarray experiments were analyzed to find biological processes that are targets of Ras-mediated signal transduction, and which are specifically regulated by Erk-activity. As the specific transcription factors that regulate a certain immediate-early gene are only known anecdotally, several groups try to find these links by bioinformatics techniques. However, the results are often misleading as they are dominated by false positives.

In this chapter a functional view on the *in silico* prediction of transcriptional regulation is proposed to circumvent the problem of false positives. A method to predict biological functions regulated by a combinatorial interaction of transcription factors is presented. Using the rigorous statistics developed in the preceding chapter, this approach intersects the presence of transcription factor binding sites in gene upstream sequences with Gene Ontology terms associated with these genes. First, profiles of transcription factors that are controlled by Ras-mediated signal transduction are discussed. Subsequently, a set of promoter found to control the expression of RANTES upon lipopolysaccharide stimulation is used to predicts functional targets of this stimulus, which are in good agreement with microarray data. Finally, the specificity and sensitivity of the method are estimated.

7.1   Introduction

The regulation of transcription is a major mechanism controlling the spatial and temporal activity of genes, thereby governing the organization of biological processes in eukaryotic organisms. The complex signaling machinery transduces external and internal stimuli to the activities of transcription factors which are the major means of transcriptional regulation. Activation of Erk, for example, controls the activity of the transcription factors CREB, Elk-1, SRF (Davis et al., 2000; Rivera et al., 1993), c-Myc, and c-Jun and c-Fos which form the AP-1 complex (Davis, 1995). Single factors controlling expression in prokaryotic cells evolved into a sophisticated regulatory machinery in eukaryotes, involving the combinatorial action of transcription factors that bind to sites which can be distributed over large regions of the genome (Cawley et al., 2004; Euskirchen et al., 2004). In eukaryotes, it is rare that individual binding sites are strongly conserved, only the combinatorial action gives rise to a specific control. Therefore, understanding complex gene regulatory networks in higher organisms is difficult.

The fundamental process of the gene regulatory machinery is the binding of transcription factors to regulatory cis-sequences located in the promoter and regulatory regions of genes. Experimentally identified cis-sequences for a single transcription factor can be aligned to find the sequence recognized by the factor. Alignments of such cis-sequences yield positional frequency matrices, which are further used to model the binding affinity of the factor toward DNA sequences (Stormo, 1998). An in-vitro method, SELEX-SAGE, allows to discover potential binding sites for transcription factors in vitro and to build positional frequency matrices without knowing their target genes (Roulet et al., 2002). For numerous transcription factors corresponding frequency matrices have been constructed (Heinemeyer et al., 1998; Sandelin et al., 2004)).

These positional frequently matrices can be used to predict binding of transcription factors (Stormo, 1998; Wasserman and Sandelin, 2004). Despite advances in this field, genome-wide scans of binding sites are difficult to interpret since false, nonfunctional predictions dominate. Wasserman and Sandelin (2004) estimate that a simple search for binding sites results in only one functional site per 1000 predictions. More recent solutions of this problem take into account further biological properties, like clustering of the functional binding sites (Frith et al., 2003) and possible conservation of cis-sequences in evolution (Dieterich et al., 2003). These approaches can reduce the number of non-functional predictions by about two orders of magnitude (Wasserman and Krivan, 2003). Nevertheless, the specificity of binding site prediction is still unsatisfactory (Wasserman and Sandelin, 2004) and expensive experimental studies such as ChIP on chip experiments are necessary (Euskirchen et al., 2004; Ren et al., 2002; Martone et al., 2003).

In this chapter a novel, more function oriented, genome-wide analysis is proposed to predict the biological function of transcription factors that utilizes the growing systematic functional annotation provided by the Gene Ontology (Ashburner et al., 2000). Using HomGL (Blüthgen et al., 2004), upstream regions of genes are extracted and scanned for clusters of binding sites using the public software Cluster-Buster (Frith et al., 2003). Then the genes with a common cluster in their upstream regions are searched for statistical association with annotations from the Gene Ontology. For this purpose the GOSSIP algorithm that is described in the preceding chapter (Chapter 6) is used. It was shown in the previous chapter that the profiles are very robust with respect to "noise" in the gene group. This property allows the prediction of biological functions controlled by combinatorial action of transcription factors, as the gene-groups predicted to be regulated are dominated by false prediction (Wasserman and Sandelin, 2004).

This chapter is structured as follows. First, the framework and data-sources are described. Second, functional targets of transcription factor sets that are controlled by Ras-mediated signal transduction are studied. Then the approach is used to bridge the gap between a detailed study of the regulation of single genes and a genome-wide analysis. Fessele et al. (2001) have unveiled the transcription factors that differentially regulate the expression of the chemokine RANTES upon lipopolysaccharide stimulation in monocytes. The framework is applied to this set of transcription factors and the predicted functions are compared with profiles of publicly available microarray data. The results show a remarkable similarity, although the microarray data have been generated in a different organism and after a long stimulation, allowing also indirect regulation. Finally, the method is further analyzed to assess specificity and sensitivity.

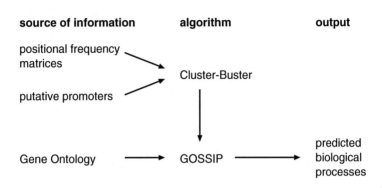

Figure 7.1: Using the Cluster-Buster algorithm, clusters of binding sites in putative promoter regions are searched. The list of genes having a cluster in their promoter is then passed to GOSSIP, which detects association with biological processes using the Gene Ontology. The significantly associated processes are reported.

7.2   The algorithms and data sources

Algorithm

The analysis presented in this chapter combines two algorithms (see Fig. 7.1). First, a genome-wide search of the genes that are potentially regulated by the factors under consideration is performed. For this purpose, the Cluster-Buster program (Frith et al., 2003) is applied to predict clusters of transcription factor binding sites in upstream regions of the genes. Cluster-Buster searches for regions of the sequence that resemble a statistical model of a motif cluster more than a model of background DNA by utilizing a hidden Markov model. To avoid problems arising from subjective parameter tuning, the default parameters of the program are used.

Second, we test whether the genes with a predicted cluster are associated with biological processes. This is performed by GOSSIP (Blüthgen et al., 2005c) using Gene Ontology annotations (Ashburner et al., 2000). The GOSSIP algorithm and the definition of the false discovery rate (FDR) are described in the preceding chapter (Chapter 6). Within this chapter, we set the threshold such that the false discovery rate is kept at 5%.

Figure 7.2: The effect of varying the length of the upstream regions on the predictions for the set of the muscle related transcription factors Mef-2, Myf and TEF (analyzed in Blüthgen et al. (2005c)). A: False Discovery Rate (FDR) of the terms *striated muscle contraction* (circles), *muscle contraction* (squares) and *muscle development* (diamonds). B: Number of genes where a cluster of binding sites has been found by the program cluster-buster annotated with the terms *striated muscle contraction* (circles), *muscle contraction* (squares) and *muscle development* (diamonds). C: Total number of genes with a cluster of binding sites predicted by Cluster-Buster. D: Number of significant terms with $FDR < 0.05$ (solid line) and $FDR < 0.01$ (dashed line).

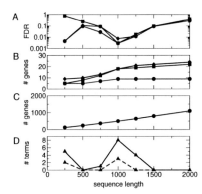

Data Preparation

For 16,032 human UniGene clusters we extracted sequences upstream of the transcription start sites reported by Ensembl (Birney et al., 2004). We found 15,362 unique upstream regions since several UniGene clusters pointed to the same genes in Ensembl. We treated the duplicates as single genes and joined their Gene Ontology annotations. We tested sequences of lengths 250, 500, 750, 1000, 1250, 1500, 2000 upstream of the TSS and found that using 1000 bp showed terms with the lowest p-values, and other lengths yielded no additional terms (see Figure 7.2). This is in agreement with the estimate by Dieterich et al. that the majority of promoters should overlap with these regions (Dieterich et al., 2002).

The Gene Ontology defines a hierarchical, controlled vocabulary to annotate genes. It contains three branches: *biological process, molecular function, cellular location*. We limited our analysis to the branch describing biological processes, as transcription is more likely to co-regulate genes involved in similar processes than molecules that share a biochemical function or a location. The annotations from the Gene Ontology were assigned to the genes using HomGL (Blüthgen et al., 2004).

In this chapter we analyze several sets of transcription factors related to Ras-mediated signal transduction. First the transcription factor sets SRF/CREB and NF-Y are represented by positional frequency matrices from the Jaspar database (Sandelin et al., 2004). Second, combinations of transcription factors that regulate the RANTES/CCL5 promoter in monocytes upon lipopolysaccharide (LPS) stimulation are analyzed (Fessele et al., 2001, 2002). This set consists of the transcription factors AP1, CEBP, CREB, ETS, NF-κB (p50 and p65), and Sp-1. It is represented by the Transfac matrices (Heinemeyer et al., 1998) with accession numbers: V\$AP1_Q6_01, V\$CEBP_Q2, V\$CREB_Q4, V\$ETS_Q4, V\$NFKAPPAB50_01, V\$NFKAPPAB65_01 and V\$SP1_Q6_01. Additionally, we evaluate the biological validity of the result for this set by analyzing a microarray data set for LPS-stimulated monocytes generated by the Alliance for Cellular Signaling, available at the signaling gateway microarray data center (`http://www.signaling-gateway.org/data/micro/cgi-bin/micro.cgi?expt=operon`), with accession numbers: MAE040216Z53, MAE040217Z53, MAE040218Z53, MAE040216Z63, MAE040217Z63 and MAE040218Z63. These are expression profiles of mouse monocytes four hours after LPS-treatment, including dye-swap and three replications. With GOSSIP we obtain biological profiles for each microarray. In this analysis, the background set consists of the significantly expressed genes ("isWellAboveBG" for both channels) and the set of regulated genes contains genes whose p-value was smaller than 0.01 ("logRatioPValue"< 0.01). Both sets are mapped to UniGene clusters with HomGL to avoid multiple entries per gene in the lists, since this could bias the analysis. The profiles for all six microarrays show identical terms associated with the up-regulated genes and none with the down-regulated genes. The datasets in the analysis of sensitivity and specificity are described in Blüthgen et al. (2005c).

7.3 Functional profiles of Erk-activated transcription factors

SRF and CREB regulate phosphatases

The two transcription factors, serum response factor (SRF) and cre-binding protein (CREB) are both activated by Erk via the activation of the ribosomal

A

B

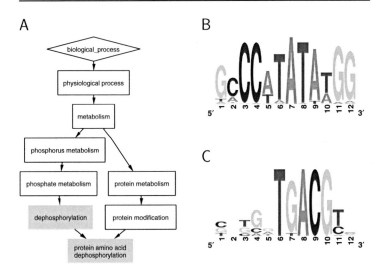

C

Figure 7.3: A: Gray boxes indicate the significantly associated biological processes (FDR≤0.05) with the transcription factors SRF and CREB. B+C: The sequence logos of these sites recognized by these factors are shown in B for SRF and C for CREB (Sandelin et al., 2004).

S6 kinases (RSK) (De Cesare et al., 1998). These transcription factors are involved in the transcriptional control of a variety of genes, but it is not known whether SRF or CREB target specific functional gene groups or pathways.

An analysis of the putative targets of both transcription factors with the method presented here shows that these two transcription factors are significantly associated with *dephosphorylation* (FDR=0.023). Interestingly, among the 309 genes that have SRF and/or CREB binding sites predicted in their upstream regions, several dual-specific phosphatases are present, such as DUSP1, DUSP2, DUSP5, and DUSP19, constituting possible negative feedbacks into the Ras-Raf-Mek-Erk signal transduction cascade. Also the phosphatase PP2A which dephosphorylates Raf is among these 309 genes. However, the phosphatase DUSP6 mediating the feedback loop that was discussed in Chapter 5 is not within this gene group. Here, the regulation is unclear, a putative binding site for NF-κB has not yet been confirmed experimentally (Tullai et al., 2004).

NF-Y regulated processes

NF-Y is a transcription factor binding the CCAAT-motive (see Fig. 7.4B). It is activated by several oncogenes including v-Ras (Gu et al., 1999). NF-Y is involved in the regulation of the G2/M transition of the cell-cycle (Lee et al., 2005), the regulation of major histocompatibility complex (MHC) class II genes (Burd et al., 2004). Therefore, the associated processes *G2/M transition of the mitotic cell cycle* and *antigen-processing via the MHC class II* are correctly predicted. Additionally, the regulation of *MHC class I*, of *S-phase*, as well as regulation of *amino-acid metabolism* are significantly associated. The involvement of NF-Y in the regulation of genes in amino-acid metabolism has been reported (Ge et al., 2002) It has been suggested recently by Jürchott et al. (2005) that NF-Y is also involved in the regulation of *S-phase of the mitotic cell cycle*. In summary, the results confirm the importance of NF-Y in cell-cycle regulation, and the regulation of MHC class II. Additionally, it suggests that NF-Y may regulate also MHC class I and proline/glutamine metabolism.

Functions regulated upon LPS stimulation

The chemokine RANTES/CCL5 plays diverse roles in the pathology of inflammatory diseases (Fessele et al., 2001). It is a chemo-attractant for T-cells and monocytes, rapidly produced in monocytes after stimulation with lipopolysaccharide (LPS). LPS is a cell-wall component of gram-negative bacteria. The presence of LPS at the outer membrane of T-cells and monocytes induces a variety of signal pathways, including Erk, JNK and IKK that in turn activate several transcription factors (see Fig. 7.5, (Alberts et al., 2002)).

Fessele et al. (2001) have investigated the transcriptional regulation of RANTES/CCL5 expression upon LPS stimulation in human monocytes. They have found that the transcription factors CREB, CEBP, NF-κB (p50/p65), Sp-1, ETS and AP1 bind the RANTES/CCL5 promoter in monocytes differently in untreated and LPS-stimulated cells. Additionally the authors have built *in silico* promoter models and have found four genes matching their model (Werner et al., 2003).

We assume that more LPS-induced genes are regulated by the same factors in monocytes. If this hypothesis holds, we could find functions that are regulated upon LPS stimulation in monocytes by applying our framework to this set of transcription factors. The resulting profile of this set of transcription factors is shown in Fig. 7.6a. We find several biological processes to be significant, including *response to stress* and *response to biotic stimulus* as well as *chemotaxis* and *cell communication*. All of them can play a role in

A

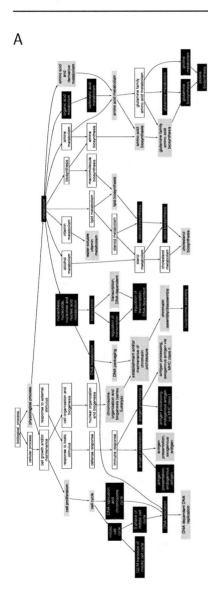

B

Figure 7.4: A: Black and gray boxes indicate the significantly associated biological processes (FDR≤0.01 and FDR≤0.05, respectively) with the transcription factor NF-Y. B: Sequence logo for the transcription factor NF-Y (Schneider and Stephens, 1990).

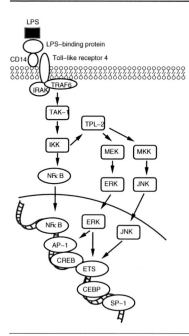

Figure 7.5: LPS-induced pathways include IKK, Erk and JNK that lead to the activation of the transcription factors NF-κB, AP-1, ETS, CREB, CEBP, Sp-1 (Fessele et al., 2001; Alberts et al., 2002; Cuschieri et al., 2004; Beinke et al., 2004).

the response of monocytes after being exposed to bacterial LPS. The terms *response to stress* and *response to biotic stimulus* are more general terms upward of inflammatory response in the hierarchy of the Gene Ontology. *Cell communication* includes the secretion of chemokines and e.g. chemokine RANTES is up-regulated upon LPS stimulation. Also a regulation of genes involved in *chemotaxis* seems plausible, since the macrophages will move toward the bacteria and secrete chemokines to attract other macrophages.

To assess the validity of the results, the predictions are compared with microarray data obtained by the Alliance for Cellular Signaling (AfCS) from mouse monocytes 4h after LPS stimulation. Profiling the list of up-regulated genes with GOSSIP yields a similar pattern of enriched GO-Terms (see Fig. 7.6b) for all six microarrays analyzed. Importantly, we find all of our predicted target functions also in the microarray data except *cell communication* and *cell adhesion*. It is not surprising that we find additional functions in the microarray data since we started our analysis from the regulation of only one gene (RANTES). It is likely that there are also other pathways and transcription factors involved in the response after LPS treatment than those which

A

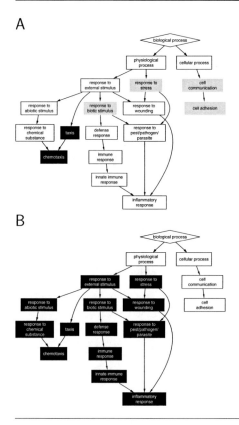

B

Figure 7.6: Black and gray boxes indicate the significantly associated biological processes (FDR≤0.01 and FDR≤0.05, respectively) with: (A) the set of transcription factors CREB, CEBP, NF-κB (p50/p65), Sp-1, ETS and AP1 as predicted by our method; (B) up-regulated genes upon lipopolysaccharide stimulation in monocytes (Microarray data from the Alliance for Cellular Signaling).

lead to the expression of RANTES. The expression profiles are expected to include the response governed by other pathways leading to more terms.

Next, investigations were made into whether the genes predicted to be regulated by the selected transcription factors are indeed up-regulated in the microarray experiment. Also, we investigated whether filtering with the significant GO-terms improves the prediction of target genes. To address this we defined three sets of genes: genes which are present on the microarray, those which have a predicted cluster of binding sites in their upstream regions, and those which are additionally involved in one of the predicted biological processes. For these three sets of genes we compute the distribution of fold-

Table 7.1: Significantly associated biological processes (FDR<0.05) for the transcription factors CREB, CEBP, p50/p65, Sp-1, ETS and AP1. (# genes: number of genes that have both a cluster of binding sites and are annotated with the GO-term; p-value: single test p-value obtained by the one sided Fisher's exact test; FDR: false discovery rate)

GO id	GO-term	# genes	p-value	FDR
6935	chemotaxis	30	$2.9 \cdot 10^5$	0.0059
42330	taxis	30	$2.9 \cdot 10^5$	0.0059
7154	cell communication	406	0.00012	0.016
6950	response to stress	120	0.00028	0.028
9607	response to biotic stimulus	131	0.00030	0.036
7155	cell adhesion	90	0.00050	0.042

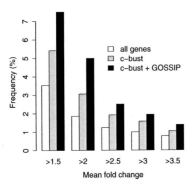

Figure 7.7: Normalized cumulative histograms of mean fold changes in the microarray data set for all genes (white bars), genes where Cluster-Buster detected a cluster of binding sites (gray bars), and after additional filtering with GO (black bars).

changes in the microarray experiment, as shown in Fig. 7.7. On the entire microarray, 216 of the 11617 genes (1.8%) have a fold change of 2 or higher. Among the 1144 genes from the microarray that have a predicted cluster of binding sites in their upstream regions, 35 genes (3%) are up-regulated (significant enrichment, $p < 0.005$ using χ^2-test). After considering these genes which are additionally annotated with the predicted GO-terms, the specificity increases further ($p < 0.02$). Out of the 360 genes which match this category, 18 genes (5%) have a fold-change higher than two. The 784 genes that do not pass the last filtering step do not differ significantly from the overall distribution ($p \approx 0.6$). These results show that a portion of the predicted target genes is up-regulated, and that the intersection with the Gene Ontology significantly improves this ratio. This confirms our initial assumption, that many genes respond to LPS through regulation by the same set of transcription factors. Additionally, it shows that the usage of functional annotations can improve the specificity of the genome-wide identification of transcription factor targets. It is particularly interesting, since our prediction has been performed on human sequences and the experiment has been done in mouse monocytes, which reflects a high degree of evolutionary conservation.

7.4 Specificity and Sensitivity

To assess the specificity of our results, we compared them to results obtained from random sets of positional frequency matrices. These random sets were constructed by permuting the positions of the matrices, thereby preserving their information content and GC-content. The analysis of 1200 sets of permuted muscle-related matrices Mef-2, Myf and TEF yielded 76 sets (6.3%) with one or more significant terms. Interestingly, none of the permuted data sets yielded results related to *muscle development* or *B-cell activation*, which are associated with the original dataset (Blüthgen et al., 2005c). On average, we found 0.46 false discoveries with $p \leq 0.0008$ (this is the p-value of the least significant term *lymphocyte differentiation*). Considering the eight significant terms in the original data set, this corresponds to an FDR of 5.8%.

For the transcription factors mediating the response of the LPS-stimulus we performed a similar analysis. 545 (27%) out of 2000 sets of permuted matrices were associated with at least one significant term and a total of 2231 terms were significant. However, care must be taken in interpreting these results, since the set contains factors recognizing sites with high GC-content (Sp-1, NF-κB p50). Although Cluster-Buster uses a background model that takes GC-variation into account, such factors prefer sites in GC-rich upstream regions. This affects the analysis since these regions themselves are associated with certain processes. The 10% of genes having the highest GC-content in their upstream regions are significantly associated with 35 terms describing processes like *development/neurogenesis, regulation of transcription, ion transport, phosphorylation* and *signal transduction* (see Fig 7.8). From the significant terms reported for the permuted matrices, 1725 (77%) were identical to the terms associated with GC-rich upstream regions, on average each of the terms occurred in 54 permuted sets. Interestingly, the terms which were not associated with GC-rich upstream regions were reported in only 1.8 sets on average. Therefore, for a correct interpretation, the composition of

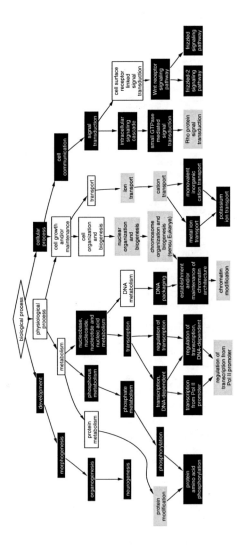

Figure 7.8: Black and gray boxes indicate the significantly associated biological processes (FDR<0.01 and 0.05, respectively) with the 10% of genes having the highest GC-content in their upstream regions.

the positional frequency matrices must be taken into account, as GC-rich matrices can induce more false positives. In such cases an analysis of permuted matrices can be used to find the expected number of false discoveries. However, the significant terms for the original set of LPS associated matrices have no significant overlap with those terms that are associated with GC-rich promoters. Furthermore, they are rarely reported in the permutation analysis: in four sets (*cell communication*), in two sets (*taxis, chemotaxis*), once (*response to biotic stimulus*), and in no set (*response to stress, cell adhesion*).

The lack of functional data for combinatorial gene regulation in higher eukaryotes makes it difficult to construct a true positive set, and, consequently, to estimate the sensitivity of our analysis. However, there are promoters with clusters of binding sites recognized by the same factor, for example clusters of E-boxes found in the promoters of circadian clock genes (Gekakis et al., 1998). These clusters can be specific enough to unveil their functional targets. For example clusters of binding sites for the transcription factor E2F can clearly be associated with the S-phase of the cell-cycle (Kielbasa et al., 2004a). Therefore, profiling of single factors might provide a rough estimate of the sensitivity. Applying our method to the 78 matrices for mammalian factors from Jaspar (Sandelin et al., 2004) and human upstream regions we found 20 significant functional profiles with 142 terms in total. As we do not know the true functions we estimate the number of false profiles by permutation analysis. Here we found on average 0.9 profiles with 4.3 terms. Given these numbers we estimate that about 19 factors out of the 78 factors under study can be correctly associated with their functional targets. Since it is not known which fraction of the 78 transcription factors exhibit clusters of binding sites in their target genes, this number cannot be translated directly into a specificity.

Many terms of the Gene Ontology, especially the more specific terms, are annotating a few genes only. For example three genes are annotated with the term *isotype switching* in the set of 15,362 unique upstream regions. In the example of the Mef-2/Myf/TEF set of transcription factors Cluster-Buster detects clusters of binding sites in 499 out of the 15,362 upstream regions. All 3 genes that describe *isotype switching* are among these 499 genes. The p-value of such a coincidence is 0.00012 (Fisher's exact test), and after multiple testing the FDR is 0.017. This demonstrates that terms that annotate only few genes can be significant. As long as the assumption that the false positive genes predicted by the cluster-buster analysis scatter randomly with respect to functional association holds the multiple-testing correction takes the number of annotations into account.

7.5 Discussion

The availability of whole-genome sequences and the growing systematic annotations like the Gene Ontology provide the means for more function oriented
data mining beyond the level of single genes. In this chapter we propose an
approach, which allows the inference of biological functions regulated by a
combinatorial interaction of transcription factors *in silico*. Contrary to other
widespread techniques our method does not intend to predict which factors
control genes of similar expression profiles. Instead our search only requires a
set of positional frequency matrices representing transcription factors to predict their biological function *in silico*. First, using Cluster-Buster we predict
a list of potential target genes for a set of transcription factors. Afterward, a
rigorous statistical test for association with biological processes implemented
in GOSSIP is applied to all biological processes provided by the Gene Ontology. Therefore the search is not biased by any prior knowledge related to
the factors and gives a chance to detect novel regulatory associations.

Our approach bridges the gap between detailed studies of single promoters and genome-wide approaches. The combinatorial action of transcription
factors found to control the gene RANTES was used to predict the regulated
functions upon lipopolysaccharide stimulation. The predicted functions show
a remarkable agreement with a profile of differentially expressed genes after
lipopolysaccharide stimulation in mouse monocytes.

Additionally the approach provides a gene list supporting the evidence of
the reported enriched processes. This gene list can be understood as the cross
section of the genes regulated by the studied factors and genes annotated with
at least one of the overrepresented terms. Due to the filtering property of the
cross section, the final gene list has less false predictions than the primary list
of potentially regulated genes, as we have validated with microarray data.

From analysis of random data sets consisting of permuted matrices we
estimate that about 5% of the terms reported to be associated with the
targets of transcription factors are chance predictions. This analysis has also
shown that the method is less specific in the case when the GC-content of
the positional frequency matrices is high, because there are several terms
associated with GC-rich upstream regions. If this is the case then the results
have to be interpreted with care, and a permutation analysis can help to
estimate the significance. Due to the lack of true positive data sets the
estimation of the sensitivity is problematic. From studying clusters of the
same binding sites for single mammalian transcription factors we estimate
that our method successfully associates biological processes with clusters of
transcription factor binding sites in more than 25% of the cases.

Several sophisticated algorithms have been developed to predict regulatory elements in higher eukaryotes. Although they use additional information like expression profiles and phylogenetic footprinting, the number of false predictions remains high. This method is not devoted to predicting the regulation of a single gene, but aims to integrate different sources of genome-wide information. It shows that with advanced prediction programs such as Cluster-Buster and the expert knowledge represented by the Gene Ontology the tools are now in hand to infer regulatory complexities when a rigorous statistic is applied. This genome-wide approach to transcription regulation allows the prediction of functions regulated by the combinatorial action of transcription factors. Additionally it can filter the list of potential target genes to reduce the number of false discoveries.

Part III

Conclusions

Erk activation is ultrasensitive

The first mathematical model of a MAPK-cascade was developed by Huang and Ferrell (1996) to investigate ultrasensitivity of Erk-activation in *Xenopus* oocytes. The mechanisms by which this early model explained ultrasensitivity were multisite phosphorylation and enzyme saturation at each level in the cascade, and multiplication of ultrasensitivity along the cascade (Ferrell and Bhatt, 1997; Brown et al., 1997). Concentrations in mammalian cells differ substantially from those in *Xenopus* oocytes such that in mammalians the enzymes in the cascade are present in comparable concentrations. Chapter 3 shows that under these conditions, enzyme saturation is an unlikely means to generate ultrasensitivity. This finding suggests that multisite phosphorylation along with amplification along the cascade is the major mechanism generating ultrasensitivity in mammalian cells. Considering the theoretical investigations of multisite phosphorylation in Chapter 2, this argument implies that ultrasensitivity of Erk-activation might be characterized by a Hill-coefficient as high as 2.8. Chapter 4 shows that the response of nuclear activated Erk upon stimulation with the platelet derived growth factor (PDGF) is ultrasensitive with a Hill-coefficient of about 4. As previous studies showed that activation of the PDGF receptor has a Hill-coefficient of 1.7 (Park et al., 2003), the stimulus-response curve of the cascade without the receptor is expected to display a Hill-coefficient above 2.4. Therefore, theoretical and experimental findings are consistent: The MAPK-cascade Raf/Mek/Erk is ultrasensitive due to multisite phosphorylation in mammalian cells with a Hill-coefficient above 2.

Ultrasensitivity in signal transduction cascades induces interesting dynamical phenomena in conjunction with feedback loops. Positive feedbacks can bring about bistability, and Bhalla et al. (2002) hypothesized that PDGF-stimulated Erk-activity in fibroblasts is bistable. This would imply that Erk-activity is all-or-none in single cells giving rise to a bi-modal distribution of Erk-activity in a population. Single-cell experiments presented in Chapter 4 show that Erk-activation is not bimodal. Thus, PDGF-induced Erk-activation is not bistable.

Kholodenko (2000) showed that only highly sensitive systems with a negative feedback might display sustained oscillations, and it requires a response coefficients of at least 8 to cause sustained oscillations. As high enzyme concentrations compromise ultrasensitivity, sustained oscillations are unlikely, and most reported data suggests that only damped oscillations are observed (Asthagiri et al., 1999b; Hornberg et al., 2004). The damped oscillations might be interpreted as adaptation or desensitization, and negative feedback loops targeting the receptor seem to be the most important signal terminator in growth-factor induced signal transduction. However, it is still unclear how growth-factor independent signaling is terminated.

Phosphatases as fundamental regulators

One possibility of signal termination is the up-regulation of phosphatases. In Chapter 5, modeling time series of phosphatase expression and Erk-activation after stimulation with oncogenic Ras suggests that DUSP6 (MKP-3), one of the most specific phosphatases of Erk, can terminate growth-factor independent signals. Although over-expressed Ras seems to outperform this negative regulator, oncogenic signals from mutated Ras expressed at physiological levels can be terminated by this signal. Thus, DUSP6 is a potential tumor suppressor.

The functional analysis of microarray data obtained for several different stimuli of the Raf/Mek/Erk cascade (such as oncogenic Ras, oncogenic Raf, and growth-factors) shows that the regulation of phosphatases is significantly associated with Erk-activation independent of the stimulus. Thus, although the functions that are regulated by Erk differ between stimuli and cell-types, the regulation of phosphatases seems to be a fundamental target. Interestingly, also predicted profiles for the Erk-regulated transcription factors SRF and CREB show exclusively dephosphorylation as the regulated function. Therefore, all three methods used in the second part of this thesis - modeling of time series, functional analysis of microarrays, and pure bioinformatic predictions of targets - highlight consistently the importance of the regulation of phosphatases by Erk.

Sequence data and annotation sufficient to unveil functions regulated by transcription factors

The major task of Erk is the regulation of transcription factors. It becomes clear that Erk-activated transcription factors are not very specific with respect to the functions of their target genes. Cross-talk to other cascades is needed to confer specificity to the regulated functions such that a specific pattern of transcription factors yields a specific result. In Chapter 7, a method that needs only sequence data, annotation of genes and descriptions of binding sites is presented, which might help to understand how specificity arises from a pattern of transcription factors. On an example of LPS-induced genes it is shown that this method can partially explain the regulation of the target functions by a set of transcription factors. Thus, it seems that sequence data and annotation sufficient to predict a functional response if one knows the transcription factors that are involved.

Outlook

The results presented in this thesis are an example of the use of systems biology in uncovering the function of signal transduction cascades. Ras-mediated signal-transduction is studied by combining small-scale single-cell experiments, data sets gained from high-throughput experiments, sequence information, functional annotation and mathematical modeling. For a complex system such as the regulation of Ras targets a systems biological approach seems appropriate. The integration of the data that comes from different sources required the development of novel algorithms, such as Gossip, HomGL, and TFGossip. These algorithms need to be refined in future to include other kinds of data which allows for more specific analysis.

Additionally, further investigation of this system requires more high-quality data. To connect signal transduction and gene expression, chromatin-immunoprecipitation experiments on microarrays (ChIP-on-chip) can be utilized to localize binding sites of transcription factors. Reporter construct may be used to gain time-resolved measurements of transcriptional activity. Additionally, a database that collects known binding sites has to be established to allow easy access to available information about regulation.

Furthermore, quantitative modeling of the signal transduction cascades requires high-quality time series data. Such time-series can then be used for estimating biochemical rate constants *in vivo*, and for the selection of the correct topology of the network.

Intracellular signal transduction cascades possess the potential for integrating data, processing them, and giving rise to an adaptive response. This thesis gives a glimpse into the mechanisms: The signal transduction cascades Raf/Mek/Erk processes the growth-factor concentration nonlinearly and therefore filters out small stimuli. More quantitative single-cell measurements of kinase activation are required to understand whether non-linear signal processing is a general principle in signal transduction cascades. The regulation of phosphatases by the cascade may account for robust adaptation and signal termination. It is remarkable that the only specific response of the two Erk-activated transcription factors SRF and CREB is the induction of phosphatases. Therefore, the function of Erk may give rise to the induction of rather unspecific immediate-early genes that gain their specificity by the activation of other cascades. At the same time Erk regulates several phosphatases. These phosphatases terminate Erk-activation but may also rearrange the signaling network by regulation of phosphatases that are more specific towards other kinases. While the kinase-network is "hard-wired" and not mainly regulated by expression changes, it is modulated by the regulation of specific phosphatases. Therefore, more attention to phosphatases as signal regulators is required to understand the dynamics and specificity of signal transduction.

Part IV

Appendix

A

Additional material on ultrasensitivity and feedbacks

Hopf bifurcation in an ultrasensitive cascade with negative feedback

This section shows a derivation of Equation 2.21, that relates the onset of bifurcations with the sensitivity of the layers of the cascade and the timescales. A slightly more general equation for this question is published by Kholodenko (2000). Consider a three-step cascade, where the activated kinases K_1, K_2 and K_3 display the following kinetics:

$$\frac{\mathrm{d}[K_1]}{\mathrm{d}t} = v_1([K_1])/[K_3] - w_1([K_1]) \qquad (A.1)$$

$$\frac{\mathrm{d}[K_2]}{\mathrm{d}t} = v_2([K_2])\,[K_1] - w_2([K_2])$$

$$\frac{\mathrm{d}[K_3]}{\mathrm{d}t} = v_3([K_3])\,[K_2] - w_3([K_3])\,.$$

Kinase K_1 activates K_2, that activates K_3, which in turn influences the activation of K_1. As the sum of the concentrations of the inactive and active form of the three kinases is preserved, both the activation kinetics v_i and deactivation kinetics w_i can be written as functions of $[K_i]$. In steady state, activation and inactivation rates are equal, i.e:

$$v_1([K_1])/[K_3] = w_1([K_1]) \qquad (A.2)$$

$$v_i([K_i])\,[K_{i-1}] = w_i([K_i]) \quad \text{for} \quad i = 1, 2\,. \qquad (A.3)$$

127

From the total differential of these equations, one obtains:

$$[K_1]\, r_3^1 \;=\; \frac{v_1/[K_3]}{\frac{\partial v_1}{\partial [K_1]}/[K_3] - \frac{\partial w_1}{\partial K_1}} \tag{A.4}$$

$$[K_i]\, r_{i-1}^i \;=\; -\frac{v_i\,[K_{i-1}]}{\frac{\partial v_i}{\partial [K_i]}\,[K_{i-1}] - \frac{\partial w_i}{\partial [K_i]}} \quad \text{for} \quad i = 1,2\,. \tag{A.5}$$

The response coefficients of the layers are defined as: $r_i^j = \frac{[K_i]}{[K_j]}\frac{\mathrm{d}[K_j]}{\mathrm{d}[K_i]}$. Linearization of the system around the steady state yields:

$$\begin{pmatrix} \frac{\mathrm{d}\Delta_1}{\mathrm{d}t} \\ \frac{\mathrm{d}\Delta_2}{\mathrm{d}t} \\ \frac{\mathrm{d}\Delta_3}{\mathrm{d}t} \end{pmatrix} = \mathbf{J} \begin{pmatrix} \Delta_1 \\ \Delta_2 \\ \Delta_3 \end{pmatrix} \tag{A.6}$$

with

$$\mathbf{J} = \begin{pmatrix} \frac{\delta v_1}{\delta [K_1]}[K_3]^{-1} - \frac{\delta w_1}{[K_1]} & 0 & -\frac{v_1}{[K_3]^2} \\ v_2 & \frac{\delta v_2}{\delta [K_2]}[K_1] - \frac{\delta w_2}{[K_2]} & 0 \\ 0 & v_3 & \frac{\delta v_3}{\delta [K_3]}[K_2] - \frac{\delta w_3}{[K_3]} \end{pmatrix} \tag{A.7}$$

being the Jacobian matrix. Δ_i represent the deviation of kinase i from the steady state. Rewriting the diagonal elements using Relations A.4 and A.5 simplifies the Jacobian matrix:

$$\mathbf{J} = \begin{pmatrix} \frac{v_1}{[K_3]}\frac{1}{r_3^1[K_1]} & 0 & -\frac{v_1}{[K_3]^2} \\ v_2 & -\frac{v_2[K_1]}{r_1^2[K_2]} & 0 \\ 0 & v_3 & -\frac{v_3[K_2]}{r_2^3[K_3]} \end{pmatrix} = \begin{pmatrix} -\frac{1}{\tau_1} & 0 & -\frac{v_1}{[K_3]^2} \\ v_2 & -\frac{1}{\tau_2} & 0 \\ 0 & v_3 & -\frac{1}{\tau_3} \end{pmatrix} \tag{A.8}$$

The diagonal elements of the Jacobian provide the invers of the timescale on which small perturbation relax. Therefore they are renamed: $\frac{1}{\tau_1} = -\frac{v_1}{[K_3]}\frac{1}{r_3^1[K_1]}$, $\frac{1}{\tau_2} = \frac{v_2[K_1]}{r_1^2[K_2]}$ and $\frac{1}{\tau_3} = \frac{v_3[K_2]}{r_2^3[K_3]}$. The characteristic equation of J is given by:

$$\prod_{i=1}^{3}\left(-\frac{1}{\tau_i} - \lambda\right) - \frac{\prod_{i=1}^{3} v_i}{[K_3]^2} = 0\,, \tag{A.9}$$

where λ is an Eigenvalue of $\mathbf{J}$. Multipying the characteristic equation by $-\prod_{i=1}^{3}\tau_i$ yields

$$\prod_{i=1}^{3}(1 + \lambda\tau_i) + \frac{\prod_{i=1}^{3} v_i\tau_i}{[K_3]^2} = 0\,. \tag{A.10}$$

Using the expressions for τ_i for the second product term, the equation simplifies to

$$\prod_{i=1}^{3} (1 + \lambda\tau_i) - r_1^2 r_2^3 r_3^1 = 0. \tag{A.11}$$

Thus, the Eigenvalues depend on the overall sensitivity and the timescales of relaxation to steady in each kinase τ_i only. To simplify notation, the overall (absolute) sensitivity is introduced as $R := -r_1^2 r_2^3 r_3^1$. In the simplest case, when all timescales are equal (i.e. $\tau_i = \tau$, $i = 1\ldots3$), the Eigenvalues are given by:

$$\lambda_{1,2} = \frac{-2 + R^{1/3}}{2t} \pm i\frac{\sqrt{3}R^{1/3}}{2t} \tag{A.12}$$

$$\lambda_3 = \frac{-1 - R^{1/3}}{t}. \tag{A.13}$$

The system possesses one non-complex negative Eigenvalue, and a pair of complex Eigenvalues. Thus it exhibits oscillations. If the real part of these complex Eigenvalues is positive, the steady-state becomes unstable and the system shows sustained oscillations (such a bifurcation is called Hopf-bifurcation). In contrast, if the real part is negative, the oscillations are damped. The real part becomes positive, and sustained oscillations appear if

$$R > 8. \tag{A.14}$$

Additional material on the effect of sequestration on ultrasensitivity

B.1 Control coefficient for sequestration in a simple covalent modification cycle

In the simple substrate cycle depicted in Fig. 3.1, the total concentrations of the substrate/target ($[T_T]$), kinase ($[K_T]$), and phosphatase ($[P_T]$) are:

$$[K_T] = [TK] + [K] \tag{B.1a}$$
$$[T_T] = [TK] + [T] + [T^*] + [T^*P] \tag{B.1b}$$
$$[P_T] = [T^*P] + [P] \tag{B.1c}$$

The derivatives with respect to the total kinase concentration are:

$$\frac{\mathrm{d}[K_T]}{\mathrm{d}[K_T]} = \frac{\mathrm{d}[TK]}{\mathrm{d}[K_T]} + \frac{\mathrm{d}[K]}{\mathrm{d}[K_T]} \tag{B.2a}$$

$$\frac{\mathrm{d}[T_T]}{\mathrm{d}[K_T]} = \frac{\mathrm{d}[TK]}{\mathrm{d}[K_T]} + \frac{\mathrm{d}[T]}{\mathrm{d}[K_T]} + \frac{\mathrm{d}[T^*]}{\mathrm{d}[K_T]} + \frac{\mathrm{d}[T^*P]}{\mathrm{d}[K_T]} \tag{B.2b}$$

$$\frac{\mathrm{d}[P_T]}{\mathrm{d}[K_T]} = \frac{\mathrm{d}[T^*P]}{\mathrm{d}[K_T]} + \frac{\mathrm{d}[P]}{\mathrm{d}[K_T]} \tag{B.2c}$$

Since the concentrations $[T_T]$ and $[P_T]$ are conserved (at least they do not vary on timescales of signal transduction), their derivatives vanish (i.e. $\frac{\mathrm{d}[T_T]}{\mathrm{d}[K_T]} = 0$ and $\frac{\mathrm{d}[P_T]}{\mathrm{d}[K_T]} = 0$). By multiplying the whole equation by $[K_T]$ these equations can be written in the notation of MCA as:

$$[K_T] = [TK]R_{K_T}^{TK} + [K]R_{K_T}^{K} \tag{B.3a}$$

$$0 = [TK]R_{K_T}^{TK} + [T]R_{K_T}^{T} + [T^*]R_{K_T}^{T^*} + [T^*P]R_{K_T}^{T^*P} \tag{B.3b}$$

$$0 = [T^*P]R_{K_T}^{T^*P} + [P]R_{K_T}^{P} \tag{B.3c}$$

These are three relations for the response coefficients. Another three relations can be obtained from the equations for the reaction rates. This systems has four reactions:

$$v_{1a}: \quad K + T \quad \leftrightarrow TK \tag{B.4a}$$

$$v_{1b}: \quad TK \quad \rightarrow T^* + K \tag{B.4b}$$

$$v_{2a}: \quad P + T^* \quad \leftrightarrow T^*P \tag{B.4c}$$

$$v_{2b}: \quad T^*P \quad \rightarrow T + P \tag{B.4d}$$

The rates of these four reactions are:

$$v_{1a} = k_{1a,f}[T][K] - k_{1a,r}[TK] \tag{B.5a}$$

$$v_{1b} = k_{1b,f}[TK] \tag{B.5b}$$

$$v_{2a} = k_{2a,f}[T^*][P] - k_{2a,r}[T^*P] \tag{B.5c}$$

$$v_{2b} = k_{2b,f}[T^*P] \tag{B.5d}$$

Differentiation these with respect to the total kinase concentration yields:

$$\frac{\mathrm{d}v_{1a}}{\mathrm{d}[K_T]} = \frac{\partial v_{1a}}{\partial[T]}\frac{\mathrm{d}[T]}{\mathrm{d}[K_T]} + \frac{\partial v_{1a}}{\partial[K]}\frac{\mathrm{d}[K]}{\mathrm{d}[K_T]} + \frac{\partial v_{1a}}{\partial[TK]}\frac{\mathrm{d}[TK]}{\mathrm{d}[K_T]}$$

$$= k_{1a,f}[K][T]\left(\frac{1}{[T]}\frac{\mathrm{d}[T]}{\mathrm{d}[K_T]} + \frac{1}{[K]}\frac{\mathrm{d}[K]}{\mathrm{d}[K_T]}\right) \tag{B.6a}$$

$$-k_{1a,r}[TK]\frac{1}{[TK]}\frac{\mathrm{d}[TK]}{\mathrm{d}[K_T]}$$

$$\frac{\mathrm{d}v_{1b}}{\mathrm{d}[K_T]} = \frac{v_{1b}}{[TK]}\frac{\mathrm{d}[TK]}{\mathrm{d}[K_T]} \tag{B.6b}$$

$$\frac{\mathrm{d}v_{2a}}{\mathrm{d}[K_T]} = \frac{\partial v_{2a}}{\partial[T^*]}\frac{\mathrm{d}[T^*]}{\mathrm{d}[K_T]} + \frac{\partial v_{2a}}{\partial[P]}\frac{\mathrm{d}[P]}{\mathrm{d}[K_T]} + \frac{\partial v_{2a}}{\partial[T^*P]}\frac{\mathrm{d}[T^*P]}{\mathrm{d}[K_T]}$$

$$= k_{2a,f}[P][T^*]\left(\frac{1}{[T^*]}\frac{\mathrm{d}[T^*]}{\mathrm{d}[K_T]} + \frac{1}{[P]}\frac{\mathrm{d}[P]}{\mathrm{d}[K_T]}\right) \tag{B.6c}$$

$$-k_{2a,r}[T^*P]\frac{1}{[T^*P]}\frac{\mathrm{d}[T^*P]}{\mathrm{d}[K_T]}$$

$$\frac{\mathrm{d}v_{2b}}{\mathrm{d}[K_T]} = \frac{v_{2b}}{[T^*P]}\frac{\mathrm{d}[T^*P]}{\mathrm{d}[K_T]} \tag{B.6d}$$

Scaling all of these by $\frac{[K_T]}{v_x}$ yields:

$$\frac{[K_T]}{v_{1a}}\frac{dv_{1a}}{d[K_T]} = \frac{k_{1a,f}[K][T]\left(R^T_{K_T}+R^K_{K_T}\right)-k_{1a,r}[TK]R^{TK}_{K_T}}{v_{1a}} \quad (B.7a)$$

$$\frac{[K_T]}{v_{1b}}\frac{dv_{1b}}{d[K_T]} = R^{TK}_{K_T} \quad (B.7b)$$

$$\frac{[K_T]}{v_{2a}}\frac{dv_{2a}}{d[K_T]} = \frac{k_{2a,f}[P][T^*]\left(R^{T^*}_{K_T}+R^P_{K_T}\right)-k_{2a,r}[T^*P]R^{T^*P}_{K_T}}{v_{2a}} \quad (B.7c)$$

$$\frac{[K_T]}{v_{2b}}\frac{dv_{2b}}{d[K_T]} = R^{T^*P}_{K_T} \quad (B.7d)$$

Since the left hand sides of Eqns. B.7a-d have to be equal (as $v_{1a} = v_{1b} = v_{2a} = v_{2b}$ in steady state), these equations yield another 3 relations for the response coefficients:

$$\text{Eqn. } B.7a \;\; = \;\; \text{Eqn.} B.7b \quad\quad\quad\quad\quad\quad\quad\quad (B.8a)$$
$$\rightarrow \; \frac{k_{1a,f}[K][T]}{v_{1a}}(R^K_{K_T}+R^T_{K_T})-\frac{k_{1a,r}[KT]}{v_{1a}}R^{TK}_{K_T}=R^{TK}_{K_T}$$

$$\text{Eqn. } B.7c \;\; = \;\; \text{Eqn.} B.7d \quad\quad\quad\quad\quad\quad\quad\quad (B.8b)$$
$$\rightarrow \; \frac{k_{2a,f}[P][T^*]}{v_{2a}}(R^P_{K_T}+R^{T^*}_{K_T})-\frac{k_{2a,r}[T^*P]}{v_{2a}}R^{T^*P}_{K_T}=R^{T^*P}_{K_T}$$

$$\text{Eqn. } B.7b \;\; = \;\; \text{Eqn.} B.7d \quad\quad\quad\quad\quad\quad\quad\quad (B.8c)$$
$$\rightarrow \; R^{T^*P}_{K_T}=R^{TK}_{K_T}$$

Equations B.3a-c and B.8a-c are six independent linear relations for the six unknown response coefficients: $R^{T^*}_{K_T}$, $R^T_{K_T}$, $R^{TK}_{K_T}$, $R^{T^*P}_{K_T}$, $R^K_{K_T}$, and $R^P_{K_T}$. Solving the equations for $R^{T^*}_{K_T}$ yields:

$$R^{T^*}_{K_T} = \frac{[K_T][P_T][T]}{[K][T^*][T^*P]+[K][P][T_T]+[P][T][TK]} \quad (B.9)$$

This equation can be further simplified, if one introduces the elasticities $\epsilon^{v_1}_T$ and $\epsilon^{v_2}_{T^*}$ of the over-all reactions v_1 and v_2 for Michaelis-Menten kinetics:

$$\epsilon^{v_1}_T = \frac{K_{M1}}{K_{M1}+[T]} = 1 - \frac{[T]}{K_{M1}+[T]} = 1 - \frac{[TK]}{[K_T]} = \frac{[K]}{[K_T]} \quad (B.10)$$

$$\epsilon^{v_2}_{T^*} = \frac{K_{M2}}{K_{M2}+[T^*]} = 1 - \frac{[T^*]}{K_{M2}+[T^*]} = 1 - \frac{[T^*P]}{[P_T]} = \frac{[P]}{[P_T]} \quad (B.11)$$

Using these elasticities, and using the mole fractions, the response coefficient can be rearranged to:

$$R_{K_T}^{T^*} = \frac{\frac{[T]}{[T_T]}}{\frac{[K]}{[K_T]}\frac{[T^*]}{[T_T]}\frac{[T^*P]}{[P_T]} + \frac{[P]}{[P_T]}\frac{[T]}{[T_T]}\frac{[TK]}{[K_T]} + \frac{[K]}{[K_T]}\frac{[P]}{[P_T]}} \tag{B.12}$$

$$= \frac{\frac{[T]}{[T_T]}}{\epsilon_T^{v_1}\frac{[T^*]}{[T_T]}\frac{[T^*P]}{[P_T]} + \epsilon_{T^*}^{v_2}\frac{[T]}{[T_T]}\frac{[TK]}{[K_T]} + \epsilon_T^{v_1}\epsilon_{T^*}^{v_2}} \tag{B.13}$$

$$= \frac{\frac{[T]}{[T_T]}}{\epsilon_T^{v_1}\frac{[T^*]}{[T_T]} + \epsilon_{T^*}^{v_2}\frac{[T]}{[T_T]} + \epsilon_T^{v_1}\epsilon_{T^*}^{v_2}\left(\frac{[TK]+[T^*P]}{[T_T]}\right)} \tag{B.14}$$

The other response coefficients can be calculated accordingly. Goldbeter and Koshland argue that the complex $[T^*P]$ might also exhibit catalytic activity, if the catalytic site and the phosphorylation site are at different positions. However, the response coefficient $R_{K_T}^{T^*+T^*P} = ([T^*]R_{K_T}^{T^*} + [T^*P]R_{K_T}^{T^*P})/([T^*P] + [T^*])$ is given by:

$$R_{K_T}^{T^*+T^*P} = \frac{[T^*][T^*P] + [T^*][P] + [T^*P][P]}{[T^*][T^*P] + [T^*][P] + [T^*P][P] + [T^*P]^2} R_{K_T}^{T^*} \tag{B.15}$$

As the denominator of the fraction before $R_{K_T}^{T^*}$ always exceeds the numerator, it is clear that the response coefficient $R_{K_T}^{T^*+T^*P}$ is smaller than $R_{K_T}^{T^*}$. Thus assuming the complex $[T^*P]$ to be active does not rescue ultrasensitivity, it even does attenuate the sensitivity coefficient.

B.2 Amount of sequestration in covalent modification cycles

As descibed in the main text, for a cycle with sequestration, where the enzymes have the same concentration and the same kinetic constants, the following relation holds:

$$2\frac{[K_T][T]}{K_M + [T]} = [T_T] - 2[T] \tag{B.16}$$

Multiplying both sides with $K_M + [T]$ yields:

$$[T]^2 + \left([K_T] + K_M - \frac{[T_T]}{2}\right)[T] - \frac{K_M}{2}[T_T] = 0 \tag{B.17}$$

Solving this equation for $[T]$ yields:

$$[T] = \frac{1}{2}\left(\frac{[T_T]}{2} - [K_T] - K_M \pm \sqrt{2K_M[T_T] + \left(\frac{[T_T]}{2} - [K_T] - K_M\right)^2}\right)$$

(B.18)

As the term in the square-root is larger than the term in front, the positive root has to be taken to ensure that $[T]$ is positive. Next, we search for the concentration of $[K_T]$ where $[T] = K_M$:

$$2K_M = \frac{[T_T]}{2} - [K_T] - K_M + \sqrt{2K_M[T_T] + \left(\frac{[T_T]}{2} - [K_T] - K_M\right)^2}$$

(B.19)

Solving this for $[K_T]$ yields:

$$[K_T] = [T_T] - 2K_M .$$

(B.20)

As long as $K_M < \frac{[T_T]}{2}$, the concentration of each species of the substrate $[T]$ is below the K_M value, if $[K_T] > [T_T] - 2K_M$.

For different catalytic activities of the kinase and phosphatase, V_1 and V_2, the amount of free target $2[T]$ for an equal amount of phosphorylated and un-phosphorylated target can be obtained by solving:

$$\left(1 + \frac{V_1}{V_2}\right)\frac{[K_T][T]}{K_M + [T]} = [T_T] - 2[T] .$$

(B.21)

Solving this equation for $[K_T]$ and summing $[K_T]$ and $[P_T] = \frac{V_1}{V_2}[K_T]$ yields:

$$[K_T] + [P_T] = 2[T_T] - 4\,K_M$$

(B.22)

Thus the concentration of both species is below the corresponding K_M values, as long as $[K_T] + [P_T] < 2[T_T] - 4\,K_M$. Therefore, higher catalytic activity of the phosphatase may partially prevent sequestration, but as soon as the sum of kinase and phosphatase concentration exceeds double the target concentration, sequestration will cause free target concentration to drop below the K_M value, when the stimulus is around the threshold (i.e. approximatly if $[T] = [T^*]$.

B.3 Models of double-phosphorylation and sequestration

This model is used to produce Fig. 3.3B of the main text:

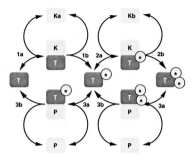

Figure B.1: The model of double phosphorylation that was used to produce Figure 3B in the main text. The phosphatase pool P is the same for dephosphorylation of T^* and T^{**}. Both reactions have the same kinetic constants, thus they are assigned the same numbers. The kinase pool, hoever, is shared for reactions 1 and 2, to simulate competition for the kinase, and was set to separate pools to simulate independent binding sites on the kinase.

$$dTK/dt = k_{1a,f}[T][Ka] - (k_{1a,r} + k_{1b,f})[TK] \tag{B.23}$$
$$dT^*/dt = k_{1b,f}[TK] + k_{3a,r}[T^*P] - k_{3a,f}[T^*][P]$$
$$-k_{2a,f}[T^*][Kb] + k_{2a,r}[T^*K] + k_{3b,f}[T^{**}P] \tag{B.24}$$
$$dT^*P/dt = k_{3a,f}[T^*][P] - (k_{3a,r} + k_{3b,f})[T^*P] \tag{B.25}$$
$$dT^*K/dt = k_{2a,f}[T^*][Kb] - (k_{2a,r} + k_{2b,f})[T^*K] \tag{B.26}$$
$$dT^{**}/dt = k_{2b,f}[T^*K] + k_{3a,r}[T^{**}P] - k_{3a,f}[T^{**}][P] \tag{B.27}$$
$$dT^{**}K/dt = k_{3a,f}[T^{**}][P] - (k_{3a,r} + k_{3b,f})[T^{**}P] \tag{B.28}$$

The model to produce the straight line in Fig. 3.3B assumes that the kinases for both phosphorylation sites is shared, thus the following relations hold:

$$[T] = T_T - [TK] - [T^*] - [T^*P] - [T^*K] - [T^{**}] - [T^{**}P] \tag{B.29}$$
$$[Ka] = [Kb] = K_T - [TK] - [T^*K] \tag{B.30}$$
$$[P] = P_T - [T^*P] - [T^{**}P] \tag{B.31}$$
$$\tag{B.32}$$

The model to produce the dotted line in Fig. 3.3B assumes that the kinases for both phosphorylation sites are not shared or have two independent docking sites that do not influence each other:

$$
\begin{aligned}
[T] &= T_T - [TK] - [T^*] - [T^*P] - [T^*K] - [T^{**}] - [T^{**}P] \quad \text{(B.33)} \\
[Ka] &= K_T - [TK] \quad \text{(B.34)} \\
[Kb] &= K_T - [T^*K] \quad \text{(B.35)} \\
[P] &= P_T - [T^*P] - [T^{**}P] \quad \text{(B.36)} \\
&\qquad\qquad\qquad\qquad\qquad\qquad \text{(B.37)}
\end{aligned}
$$

Parameter values: $k_{1a,f} = 5$, $k_{1a,r} = 4$, $k_{1b,f} = 1$, $k_{2a,f} = 0.5$, $k_{2a,r} = 40$, $k_{2b,f} = 10$, $k_{3a,f} = 0.5$, $k_{3a,r} = 40$, $k_{3b,f} = 10$, $T_T = 100$, $K_T = 30$, $P_T = 5$.

B.4 Model of MAPK cascade with sequestration and negative feedback

KKK, KK and K are MAPKKK, MAPKK and MAPK, respectively. An appended P, and PP marks phosphorylation and double-phosphorylation. The symbol $\_$ marks complexes. The model structure (without feedback, i.e. $k_{loop} \to \infty$) corresponds to the model according to Huang & Ferrell (1996).

Parameters

Parameter values are derived from the model of Kholodenko (2000). Enzyme catalyzed reactions are broken down into elementary steps, where the enzyme binds to the substrate reversibly, and modifies and releases it irreversibly:

$$
E + S \underset{d}{\overset{a}{\underset{\leftarrow}{\rightarrow}}} ES \overset{V}{\rightarrow} E + P \quad \text{(B.38)}
$$

V_i values are the catalytic constant and correspond directly to Kholodenko's V_i values for kinase reactions and are scaled by the phosphatase concentration for phosphatase reactions. a_i and d_i constants are calculated from the V_i and Kholodenko's K_M values by assuming that $d_i = 4 V_i$, i.e. $a_i = 5 V_i / K_{Mi}$ and $d_i = 4 V_i$. The ode files (model definition files for xpp-auto) used for the calculations can be obtained from the authors upon request. The following table summarizes the reaction parameters used. K_M-values are given only for reference to the original model.

no.	V	a	d	K_M-value
1	$V_1 = 2.5$	$a_1 = 1.25$	$d_1 = 10$	$K_{M,1} = 10$
2	$V_2 = 0.25/E2tot$	$a_2 = 0.15625/E2tot$	$d_2 = 1/E2tot$	$K_{M,2} = 8$
3	$V_3 = 0.025$	$a_3 = 0.00833$	$d_3 = 0.1$	$K_{M,3} = 15$
4	$V_4 = 0.025$	$a_4 = 0.00833$	$d_4 = 0.1$	$K_{M,4} = 15$
5	$V_5 = 0.75/[KKStot]$	$a_5 = 0.25/[KKStot]$	$d_5 = 3/[KKStot]$	$K_{M,5} = 15$
6	$V_6 = 0.75/[KKStot]$	$a_6 = 0.25/[KKStot]$	$d_6 = 3/[KKStot]$	$K_{M,6} = 15$
7	$V_7 = 0.025$	$a_7 = 0.00833$	$d_7 = 0.1$	$K_{M,7} = 15$
8	$V_8 = 0.025$	$a_8 = 0.00833$	$d_8 = 0.1$	$K_{M,8} = 15$
9	$V_9 = 0.5/[KStot]$	$a_9 = 0.166/[KStot]$	$d_9 = 2/[KStot]$	$K_{M,9} = 15$
10	$V_{10} = 0.5/[KStot]$	$a_{10} = 0.166/[KStot]$	$d_{10} = 2/[KStot]$	$K_{M,10} = 15$

The default value of parameter k_{loop} is 9. The total concentrations are taken from Kholodenko (2000) for the kinases. Total concentrations of the phosphatases are taken as a fraction $f > 0$ of the corresponding kinases's concentration, i.e:

concentration	value
$E1tot$	1
$E2tot$	1
$KKKtot$	100
$KKtot$	300
$Ktot$	300
$KKStot$	$300\,f$
$KStot$	$300\,f$

Dynamics of MAPKKK

$$dKKKP/dt = V_1[KKK\_E1] + d_2[KKKP\_E2] \quad (B.39)$$
$$-a_2[KKKP][E_2]$$
$$-a_3[KK][KKKP] + (d_3 + V_3)[KK\_KKKP]$$
$$-a_4[KKP][KKKP] + (V_4 + d_4)[KKP\_KKKP]$$

$$dKKK\_E1/dt = \frac{a_1[KKK][E1]}{1 + KPP/k_{loop}} - (d1 + V1)[KKK\_E1] \quad (B.40)$$

$$dKKKP\_E2/dt = a_2[KKKP][E2] - (d_2 + V_2)[KKKPE2] \quad (B.41)$$

$$[E2] = E2tot - [KKKP\_E2] \quad (B.42)$$

$$[E1] = E1tot - [KKK\_E1] \quad (B.43)$$

Dynamics of MAPKK

$$dKK\_KKKP/dt = a_3[KK][KKKP] \qquad\qquad (B.44)$$
$$-(d_3 + V_3)[KK\_KKKP] \qquad (B.45)$$
$$dKKP/dt = V_3[KK\_KKKP] + V_6[KKPP\_KKS] \quad (B.46)$$
$$+d_4[KKP\_KKKP] + d_5[KKP\_KKS]$$
$$-a_4[KKP][KKKP] - a_5[KKP][KKS]$$
$$dKKP\_KKKP/dt = a_4[KKP][KKKP] \qquad\qquad (B.47)$$
$$-(V_4 + d_4)[KKP\_KKKP]$$
$$dKKPP/dt = V_4[KKP\_KKKP] \qquad\qquad (B.48)$$
$$+d_6[KKPP\_KKS] \qquad\qquad (B.49)$$
$$-a_6[KKPP][KKS] - a_8[KP][KKPP]$$
$$(V_8 + d_8)[KP\_KKPP] - a_7[K][KKPP]$$
$$+(V_7 + d_7)[K\_KKPP]$$
$$dKKPP\_KKS/dt = a_6[KKPP][KKS] \qquad\qquad (B.50)$$
$$-(V_6 + d_6)[KKPP\_KKS]$$
$$dKKP\_KKS/dt = a_5[KKP][KKS] - (V_5 + d_5)[KKP\_KKS] \quad (B.51)$$
$$KKS = KKStot - KKPP\_KKS - KKP\_KKS \quad (B.52)$$

Dynamics of MAPK

$$dK\_KKPP/dt = a_7[K][KKPP] - (V_7 + d_7)[K\_KKPP] \tag{B.53}$$
$$dKP/dt = V_7[K\_KKPP] + V_{10}[KPP\_KS] \tag{B.54}$$
$$+ d_8[KP\_KKPP] + d_9[KP\_KS]$$
$$- a_8[KP][KKPP] - a_9[KP][KS] \tag{B.55}$$
$$dKP\_KKPP/dt = a_8[KP][KKPP] - (V_8 + d_8)[KP\_KKPP] \tag{B.56}$$
$$dKPP/dt = V_8[KP\_KKPP] + d_{10}[KPP\_KS] \tag{B.57}$$
$$- a_{10}[KPP][KS]$$
$$dKPP\_KS/dt = a_{10}[KPP][KS] - (d_{10} \tag{B.58}$$
$$+ V_{10})[KPP\_KS]$$
$$dKP\_KS/dt = a_9[KP][KS] - (d_9 + V_9)[KP\_KS] \tag{B.59}$$
$$KS = KStot - [KPP\_KS] - [KP\_KS] \tag{B.60}$$
$$KKK = KKKtot - [KKKP] - [KKKE1] \tag{B.61}$$
$$- [KKKPE2] - [KK\_KKKP] - [KKP\_KKKP]$$
$$KK = KKtot - [KKP] - [KK\_KKKP] \tag{B.62}$$
$$- [KKP\_KKKP] - [KKPP] - [KKPP\_KKS]$$
$$- [KKP\_KKS] - [K\_KKPP] - [KP\_KKPP]$$
$$K = Ktot - [KP] - [KPP] - [K\_KKPP] \tag{B.63}$$
$$- [KP\_KKPP] - [KPP\_KS] - [KP\_KS]$$

C

Quantification of Erk-activity in single cells

As a measure of Erk-activity, the intensity of the immunofluorescence in the nuclei was used. From each area of the slides two images were taken: an image for the DAPI-staining that marks areas with DNA (mainly the nucleus), and an image that shows the immunofluorescence of anti-pErk antibodies. The image analysis performed on these images had to first determine the areas of the nuclei and then to quantify the phosphor-Erk immunofluorescence in a way that makes the samples comparable, reduce the background and reduces effects due to non-homogeneous illumination.

Finding the nuclei

The contrast between the DAPI-signal and the background was rather high. After applying a threshold of 1.4 times the median, most of the remaining pixels belong to the areas of the nuclei. A further clustering of these pixels (grouping neighbouring pixels) was used to define connected areas of high intensity. Afterwords only these areas were used that contained more than 300 pixels. These clusters were then visualized and visually inspected to remove clusters that do not represent a nucleus.

Quantification of Erk-activity

As a first model, the mean intensity of the pixels in the areas of the nucleus from the immunofluorescence image were taken and the median of the immunofluorescence image was subtracted as the background, since most of the space in the images was not occupied by cells and can be regarded as

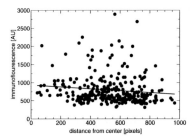

Figure C.1: The quantified immunofluorescence in the nucleus using a naive background model (subtracting median intensity) showed a negative correlation with the distance from the center. the correlation coefficient is R=-0.12, and the slope of the least square fit is -0.26.

background:

$$P(x, y) = b + I(x, y) + \xi \quad , \tag{C.1}$$

with $P(xm, y)$ being the observed pixel-value, b being the global background and ξ being some noise. However, quantification of the immunofluorescence in the nuclei unveiled a dependence of the immunofluorescence on the distance from the center of the image (see Fig. C.1). A possible reason for this is that the illumination of the microscope was not homogeneous. To reflect this inhomogeneous illumination we used a more elaborated model that includes a multiplicative term $\Lambda(x, y)$:

$$P(x, y) = \Lambda(x, y) \left(b + I(x, y) \right) + \xi \tag{C.2}$$

Estimation of the multiplicative term $\Lambda(x, y)$ using two-dimensional locally weighted scatter plot smooth (LOWESS) was too slow for processing all images, and approximation of $\Lambda(x, y)$ using one-dimensional LOWESS did only reduce the slope of the correlation by about a factor two.

More successful was a parametric model for $\Lambda(x, y)$:

$$b\Lambda(x, y) = p_1 - \frac{(x - x_0)^2 + (y - y_0)^2}{p_1} \quad . \tag{C.3}$$

Here a least trimmed sum of squares (LTS) regression was applied to estimate the parameters. Least trimmed sum of squares minimizes the sum of squares but takes only a certain quantile of data points into account, in this case only the 60% of the datapoints that lay closest to the fit. This reflects the

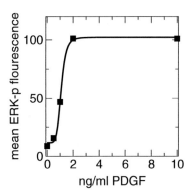

Figure C.2: Stimulus response curve of mean Erk-activity as a function of PDGF stimulation has a Hill coefficient of 5.9 as evaluated by least-sqare fit with a Hill curve. Mean activity was read of Fig. 3C of MacKeigan et al. (2005) and calculated as described in this appendix.

idea that large areas of the image are pure background and that fitting of $b\Lambda(x.y)$ to 60% of the pixels will fit only the background. The normalization factor b was calculated by averaging over $\Lambda(x, y)$:

$$b \approx \frac{1}{x_{max}\, y_{max}} \int_0^{x_{max}} \int_0^{y_{max}} p_1 - \frac{(x - x_0)^2 + (y - y_0)^2}{p_1}\, dy\, dx \quad \text{(C.4)}$$

$$= p_1 - \frac{1}{p_2}\left(x_0^2 - x_0 x_m + \frac{1}{3}x_m^2 + y_0^2 - y_0 y_m + \frac{1}{3}y_m^2 \right) \quad . \quad \text{(C.5)}$$

Comparison with Data by MacKeigan et at.

MacKeigan et al. (2005) performed a similar experiment as we did. They stimulated mouse fibroblasts (swiss 3T3 cells) with different PDGF concentrations and quantified ERK-activity by immuno-staining and subsequent FACS analysis. As their histograms are shown in log-scale, the histograms are not directly comparable. As the distributions seem to be normal on log-scale, means could not directly been read-off by reading of the peak of the distribution. However, taken peak and width of the distribution, one can calculate the mean value:

$$
\frac{<x>}{x_0} = \frac{\int_0^\infty x\, e^{-(\log(x)-\log(x_0))^2/2\sigma^2}}{x_0 \int_0^\infty e^{-(\log(x)-\log(x_0))^2/2\sigma^2}}
$$

$$
= e^{\frac{3}{4}2\sigma^2} \tag{C.6}
$$

Using this formula, the maxima and standard deviations estimated from Fig. 3C of MacKeigan et al. (2005) were used to plot the stimulus-response curve in Fig. C.2. Best fit of this stimulus-response curve was obtained with a Hill coefficient of 5.9. Thus the data by MacKeigan et al. (2005) yields a similar conclusion: Erk-activation by PDGF is ultrasensitive. Additionally, also their data shows higher variation in Erk-activity for cells stimulated with PDGF concentrations near the threshold.

D

Family-wise error rate (FWER) of functional profiles

This chapter covers additional topics of functional profiling of gene groups.

If there is no a priori expectation of any term to be enriched in the test group, controlling the family-wise error rate (FWER) is regarded as the appropriate multiple testing correction Dudoit et al. (2003). The FWER provides the probability to get *any* false discovery:

$$FWER(\alpha) = Pr(NFD(\alpha) > 0) \,. \tag{D.1}$$

Here $NFD(\alpha)$ denotes the number of false discoveries given a threshold of α for the single-test p-values. If all tests were independent, we could calculate the family-wise error rate for the list of terms from term $t = 1$ to term $t = i$ iteratively by

$$
\begin{aligned}
FWER_i(\alpha) &= FWER_{i-1}(\alpha) \\
&\quad + Pr(p_i \leq \alpha)(1 - FWER_{i-1}(\alpha)) \,,
\end{aligned}
\tag{D.2}
$$

starting the iteration with $FWER_0(\alpha) = 0$, and $Pr(p_i \leq \alpha)$ being defined as above. The iteration ends at the end of the list of annotated terms and yields the family-wise error for the entire list. However, there are strong correlations between the annotation of terms due to the graph-structure and also correlations between annotations of single genes (e.g. certain functions correspond to certain cellular locations). Therefore the joint probability of term t having a p-value below the threshold *and* no p-value of any other term in the list before passing the threshold is overestimated by simple multiplication of the probabilities $Pr(p_t \leq \alpha)(1 - FWER_{t-1}(\alpha))$. We find empirically, that the joint probabilities can be very well estimated by $\beta Pr(p_i \leq \alpha)(1 - FWER_{i-1}(\alpha))$

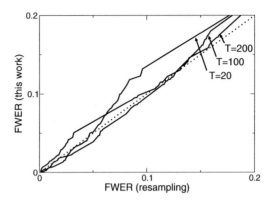

Figure D.1: Comparison of family-wise error rate (FWER) obtained by 1000 resampling runs and estimated by the heuristic method presented here. The test groups of size $T = 20$, $T = 100$ and $T = 250$ were chosen randomly from all GO-annotated probe-sets on Affymetrix's HG-133A (solid lines, 15652 annotated probe-sets). The dotted line indicates the diagonal.

with the parameter $\beta = \frac{1}{4}$. This estimate leads to the following heuristic adaptation of Equation D.2:

$$
\begin{aligned}
FWER_i(\alpha) &= FWER_{i-1}(\alpha) \\
&\quad + \beta Pr(p_i \leq \alpha)\left(1 - FWER_{i-1}(\alpha)\right).
\end{aligned} \tag{D.3}
$$

We tested this heuristic formula with a variety of data sets including different Affymetrix chips, gene groups of small customized microarrays and all annotated UniGene genes and found a good agreement of the estimated $FWER(\alpha)$ by this method with the $FWER(\alpha)$ obtained by resampling. Figure D.1 shows examples of this study. Since the $FWER(\alpha)$ is already strictly monotonic with increasing α we can utilize the $FWER(\alpha)$ as an adjusted p-value p_{FWER} controlling the probability of having any false discoveries in the list. For instance, if one chooses a threshold of $p_{FWER} \leq 0.05$, one obtains a list where the probability of having one or more falsely discovered terms is 5%.

Figure D.2 shows that our approach approximates the FWER obtained by resampling simulations very well, especially in the important region of adjusted p-values below 0.05. The other methods (Holm, Sidak, Bonferoni) are too conservative resulting in adjusted p-values that are around 2.5-6 times

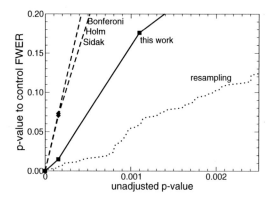

Figure D.2: Adjusted p-value to control the FWER for the group of down-regulated genes in the cerebellum in the Huntington's disease mouse model. The FWER estimated by our approach (solid) is in good agreement with the results from resampling simulations (dotted), especially in the important region of low p-values. Squares indicate the values of the adjusted p-values for the GO-terms. The adjusted p-values by Bonferoni, Holm and Sidak (dashed) are similar to each other and about 2.5-6 fold too conservative when compared to resampling simulations.

larger, and they thus miss significant results.

E

Fitting and evaluating models

This appendix should give a brief overview over the methods used to fit and select models in Chapter 5. It is inspired from Jens Timmer's script on "Mathematische Methoden zur Analyse von Zeitreihen komplexer Systeme", and stimulating discussions with Thomas Maiwald.

E.1 Maximum likelihood fit

The most commmon method to fit a model to data is to maximize the likelihood, e.g. find the parameters θ that maximize the joined probability density of the data given the model and the parameters:

$$\theta_{ML} = \underset{\theta}{\operatorname{argmax}}(P(D|M,\theta)).$$ (E.1)

$D = \{d_1,\ldots,d_N\}$, $M = \{m_1,\ldots,m_N\}$, and $\theta = \{p_1,\ldots,p_r\}$ represent the data, the model and the parameters, respectively. For the likelihood $L(\theta|D,M) := P(D|M,\theta)$ it is often assumed that the data is subject to normaly distributed and uncorrelated error.[1] Consequently, the likelihood can be written as a product of gaussian distributions:

$$L(\theta|D,M) := P(D|M,\theta) = \prod_i \frac{1}{\sqrt{2\pi}\sigma_i} e^{-\frac{(d_i-m_i)^2}{2\sigma_i^2}}.$$ (E.2)

It is often more convenient to maximize the log-likelihood $\mathcal{L}(\theta|M,D)$, since first it is a sum and not a product, and second the values of the likelihood

---

[1]P is the probability distribution of observing the data, given the model (including an error model), and a parameter set. Note that in the notation of the likelihood the dependencies of the variables are reversed.

functions are often very small:

$$\mathcal{L}(\theta|M, D) \ := \ \log L(\theta|D, M)$$
$$= \ -\sum_i \frac{(d_i - m_i)^2}{2\sigma_i^2} - N \log \sqrt{2\pi} - \sum_i \log \sigma_i \, . \quad \text{(E.3)}$$

As the last two terms in the log-likelihood function are constants that depend solely on the data and the errormodel, they can be neglected when the likelihoods are maximized or the log-likelihood-ratios between two parameter-sets or two models are computed. Thus, instead of comparing log-likelihood one can compare the χ^2 values:

$$\chi^2 \ = \ \sum_i \frac{(d_i - m_i)^2}{\sigma_i^2} \, . \quad \text{(E.4)}$$

E.2 Comparing models

Popular model selection strategies include step-up, step-down and score-based methods. In step-up one starts with a very simple model and extends the model further until no significant improvement can be detected. A step-down procedure takes a very complex model and reduces it until the fit gets significantly worse. For both strategies a likelihood ratio test can be applied. In contrast, in score-based approaches (using a score like the Akaike's information criterion, see below) several models are compared and the model with the best score is selected. In the following, likelihood ratio test and the Akaike's information criterion are described.

Likelihood ratio test

When reducing a model's complexity, one may want to ask whether the reduced model sufficiently explains the data, or whether the more complex model is needed to describe the data. In other words, one wants to test whether a more complex model M_1 fits better then a derived, less complex model M_2, and does not only improve the fit of the noise. One can compare the likelihoods of the model. Then, the null-hypothesis (H_0) is that the reduced model (M_2) is the true model, and (M_1) is a model with more parameters that produces the data equally well. In contrast, the hypothises H_1 is that the true model is model M_2, and that the model M_1 is not generating the data. To test these hypotheses, one compares the two log-likelihoods $\mathcal{L}(\theta_{1,ML}|M_1, D)$ and $\mathcal{L}(\theta_{2,ML}|M_2, D)$. As the more complex model has allways a higher likelihood, the question which log likelihood-ratio is significant.

In the following, it will be outlined that one can use χ^2-statistics, if one assumes that the maximum-likelihood parameters θ_{ML} are normaly distributed around the true parameters θ_0 with covariance matrix $\mathbf{C}$ given by the inverse of the Hesse matrix, and the parameters are identifiable[2]. Subsequently, a bootstraping procedure is introduced to estimate the p-value if the assumtions above are violated.

Log-likelihood-ratio of a model is χ^2-distributed

First it will be shown that the log-ratio between the likelihood of true parameter set and the estimated parameter set follows a χ^2-distribution. It is assumed that the θ_{ML} are normaly distributed around the true parameters θ_0 with covariance matrix $\mathbf{C}$:

$$P(\theta_{ML} - \theta_0) = \frac{1}{\sqrt{2\pi}} |\mathbf{C}|^{-\frac{1}{2}} \exp\left(-\frac{1}{2}(\theta_{ML} - \theta_0)^T \mathbf{C}^{-1}(\theta_{ML} - \theta_0) \right). \quad \text{(E.5)}$$

with

$$\mathbf{C}_{i,j} = -\left(\frac{\partial^2 \mathcal{L}(\theta_{ML}|M,D)}{\partial \theta_i \partial \theta_j} \right)^{-1}. \quad \text{(E.6)}$$

The likelihood of the estimated parameters can be approximated using Taylor-expansion,

$$\begin{aligned}
\mathcal{L}(\theta_0|M,D) &\approx \mathcal{L}(\theta_{ML}|M,D) + \frac{\partial \mathcal{L}(\theta_{ML}|M,D)}{\partial \theta}(\theta_0 - \theta_{ML}) \\
&+ \frac{1}{2}(\theta_0 - \theta_{ML})^T \frac{\partial^2 \mathcal{L}(\theta_{ML}|M,D)}{\partial \theta_i \partial \theta_j}(\theta_0 - \theta_{ML}). \quad \text{(E.7)}
\end{aligned}$$

At maximum likelihood, however, the first derivatives $\frac{\partial \mathcal{L}(\theta_{ML}|M,D)}{\partial \theta}$ equals zero. Thus the formula reduces to:

$$\mathcal{L}(\theta_0|M,D) \approx \mathcal{L}(\theta_{ML}|M,D) - (\theta_0 - \theta_{ML})^T \mathbf{C}^{-1}(\theta_0 - \theta_{ML}).$$

$$\rightarrow \quad \mathcal{L}(\theta_{ML}|M,D) - \mathcal{L}(\theta_0|M,D) \approx (\theta_0 - \theta_{ML})^T \mathbf{C}^{-1}(\theta_0 - \theta_{ML}). \quad \text{(E.8)}$$

As the covariance matrix is symmetric and positive definite, a transformation $\mathbf{U}$ exists under which the covariance matrix is diagonal, and also the transformed inverse is diagonal with

$$\mathbf{U}^T \mathbf{C}^{-1} \mathbf{U} =: \tilde{\mathbf{C}}^{-1} = \mathrm{diag}(1/v_1, 1/v_2, ..., 1/v_r). \quad \text{(E.9)}$$

---

[2]Another prerequisite is that the parameters are not at the border of the parameterspace, as the normal distribution does not hold in this case.

Then Eqn. E.8 can be reformulated as:

$$\mathcal{L}(\theta_{ML}|M, D) - \mathcal{L}(\theta_0|M, D) = (\theta_0 - \theta_{ML})^T \mathbf{U}\tilde{\mathbf{C}}^{-1}\mathbf{U}^T(\theta_0 - \theta_{ML})$$

$$= \sum_{i=1}^{r} \frac{1}{v_i}(\mathbf{U}^T(\theta_0 - \theta_{ML}))_i^2. \qquad (E.10)$$

As the transformation $\mathbf{U}$ was constructed such that the components of the transformed parameter differences $(\mathbf{U}^T(\theta_0 - \theta_{ML}))_i$ have a variance of v_i and no covarince, the log likelihood-ratio is a sum of r normalized gaussian distributed random numbers and thus χ^2 distributed with r degrees of freedom.

Comparing nested models

If one compares two models that are nested, e.g. model M_1 is a reduction of M_2, the log-ratio of their likelihoods under the null-hypothisis that M_1 is the true model follows also a χ^2-distribution. If one performs a Taylor-expansion of both $\mathcal{L}(\theta_{1,ML}|M_1, D)$ and $\mathcal{L}(\theta_{2,ML}|M_2, D)$ (comapare Eqn. E.7), the first terms of these series $\mathcal{L}(\theta_{1,0}|M_1, D)$ and $\mathcal{L}(\theta_{2,0}|M_2, D)$ are equal, as model M_2 has the same solution under the null hypotheses with the true parameters and thus explains the data equally well. The linear terms are zero as the likelihood is maximized, and only the quadratic terms remain in the log-ratio:

$$\mathcal{L}(\theta_{1,0}|M_1, D) - \mathcal{L}(\theta_{2,0}|M_2, D)$$
$$\approx (\theta_{1,0} - \theta_{1,ML})^T \mathbf{C_1}^{-1}(\theta_{1,0} - \theta_{1,ML})$$
$$- (\theta_{2,0} - \theta_{2,ML})^T \mathbf{C_2}^{-1}(\theta_{2,0} - \theta_{2,ML}). \qquad (E.11)$$

If one assumes without loss of generality[3] that the r_1 parameters of M_1 equal the first r_1 parameters of M_2, the first r_1 values in the scalar product vanish and the only remaining $r_2 - r_1$ parameters contribute the the log-likelihood-ratio.

Monte-Carlo estimation of the distribution

If the assumptions under which the log-likelihood-ratio follows a χ^2 distribution do not hold or if the models are not nested, one can generate the distribution of the log-likelihood-ratios by a Monte-Carlo procedure as follows. Under the null hypotheses, that the less complex model M_1 is the

---

[3]One may need to find a new parametrazation of where r_1 parameters of M_1 are the same for model M_2 and the remaining $r_2 - r_1$ parameters of M_2 describe the extension of M_2.

true model, this model is fitted to the data by a maximum likelihood procedure. Subsequently, this fitted model is used to generate data sets subject to some measurement noise, using the same error-model as used to calculate the likelihood. Then both models are fitted to the generated data sets and the distribution of the log-likelihood-ratios resamples the log-likelihood-ratio distribution under the null hypotheses.

Akaike's information criterion

Another method to evaluate the best, i.e. most parsimonious model is given by Akaike's information criterion (AIC).[4] It is based on information theory, and evaluates the Kullback-Leibler distance between the truth and the model (Burnham and Anderson, 1998):

$$I(P_t, P_m) = \int P_t(D) \log \left(\frac{P_t(D)}{P(D|M, \theta)} \right) \mathrm{d}D. \tag{E.12}$$

Here, $P_t(D)$ and $P(D|M, \theta)$ are the probability distributions of the data for the truth and the model. The Kullback-Leibler distance expresses the information that is "lost" when the model is used to approximate the the truth. Eqn. E.12 can be expandet to:

$$I(P_t, P_m) = \int P_t(D) \log P_t(D) \mathrm{d}D - \int P_t(D) \log P(D|M, \theta) \mathrm{d}D. \tag{E.13}$$

As the first term is constant for a given "truth", only the second term has to be evaluated to compare models:

$$\int P_t(D) \log P(D|M, \theta) \mathrm{d}D = \int P_t(D) \mathcal{L}(\theta|M, D) \mathrm{d}D. \tag{E.14}$$

The second term is given by the expectation value of the likelihood weighted with the probability distribution of the true model. Two problems arise when computing this term. First, the truth $(P_t(D))$ is not known, and second, the best parameter set is not known and just estimated by e.g. a maximum likelihood fit. Akaike (1973) found that the second term can be approximated by

$$\mathcal{L}(\theta_{ML}|M, D) - r. \tag{E.15}$$

Since usually χ^2-values are computed to fit a model, the formula was multiplied by -2:

$$AIC = -2\mathcal{L}(\theta_{ML}|M, D) + 2r = \chi^2 + 2r. \tag{E.16}$$

---

[4]Originally, it was called *an* information criterion.

With AIC, one would use the differences in the AIC to compare models, and would select the model with the lowest AIC. The detailed derivation of AIC and alternative estimators are intensively discussed in Burnham and Anderson (1998).

Abbreviations

AP-1 transcription factor complex composed of Fos and Jun

Akt v-akt murine thymoma viral oncogene homolog, a kinase

CEBP CCAAT/enhancer binding protein (C/EBP), a transcription factor

CREB cAMP responsive element binding protein, a transcription factor

ChIP chromatin immuno-precipitation

DUSP dual-specific phosphatase

EGFR EGF receptor

EGF epidermal growth factor

EST expressed sequence tag

ETS v-ets erythroblastosis virus E26 oncogene homolog, a transcription factor

Elk-1 v-ets erythroblastosis virus E26 oncogene homolog, a transcription factor

Erk extracellular-regulated kinase, a MAPK.

FDR false discovery rate

FWER family-wise error rate

Fos FBJ murine osteosarcoma viral (fos) oncogene homolog, a transcription factor

GDP guanine diphosphate

GTP guanine triphosphate

HEK human embryonic kidney

HePTP protein tyrosine phosphatase, non-receptor type

IEG immediate-early gene

IκK inhibitor of NF-κB

IPTG isopropyl-β-D-thiogalactoside

JAK Janus tyrosine kinase

Jnk Jun N-terminal kinase family member, a MAPK.

Jun jun sarcoma virus 17 oncogene homolog, a transcription factor

LPS lipopolysaccharide

MAPKKK mitogen-activated protein kinase kinase kinase, MAPKK kinase.

MAPKK mitogen-activated protein kinase kinase, MAPK kinase.

MAPK mitogen-activated protein kinase, MAP kinase.

MCA metabolic control analysis

MHC major histo-compatibility complex

MKP MAPK phosphatase, old name for DUSPs

Mef-2 Myocyte enhancing factor 2, a transcription factor

Mek MAP kinase kinase or Erk kinase, a MAPKK.

Msk Mitogen- and stress-activated kinase

Myc myc myelocytomatosis viral oncogene homolog, a transcription factor

Myf Myogenic factor, a transcription factor

NF-κb nuclear factor of kappa light chain gene enhancer

NF-Y nuclear transcription factor Y

NFAT nuclear factor of activated T-cells, a transcription factor

PC12 neuronal precursor cell line

PDGF platelet-derived growth factor

PI3K phosphatidylinositol-4-phosphate 3-kinase

PKC protein kinase C

PMA phorbol 12-myristate 13-acetate

PP2A protein phosphatase 2A

RANTES a chemokine

RB retinoblastoma interacts with cell-cycle relevant transcription factors.

RSK ribosomal protein S6 kinase, nuclear kinase

Raf v-raf-1 murine leukemia viral oncogene homolog, a MAPKKK.

Ras rat sarcoma viral oncogene homolog proteins, small GTP/GDP binding protein.

Rsk Ribosomal protein S6 kinase

SOS son of sevenless, a guanyl-nucleotide exchange factor

SP-1 SP1 transcription factor

SRF serum response factor, a transcription factor

STAT signal transducer and activator of transcription, transcription factor

TEF thyrotrophic embryonic factor, a transcription factor

TF transcription factor

cPLA$_2$ cytosolic phospholipase a2 in B-cells, transcription factor

Bibliography

Akaike, H., 1973. Information theory as an extension of the maximum likelihood principle. In: Petrov, B., Csaki, F. (eds.), Second international symposium on information theory. Akademiai Kiado, Budapest, pp. 267–281.

Al-Shahrour, F., Diaz-Uriarte, R., Dopazo, J., 2004. FatiGO: a web tool for finding significant associations of Gene Ontology terms with groups of genes. Bioinformatics 20, 578–580.

Albe, K. R., Butler, M. H., Wright, B. E., 1990. Cellular concentrations of enzymes and their substrates. J. Theor. Biol. 143, 163–195.

Alberts, B., Johnson, A., Lewis, J., Raff, M., Roberts, K., Walter, P., 2002. Molecular biology of the cell. Garland Science (New York).

Alessi, D. R., Gomez, N., Moorhead, G., Lewis, T., Keyse, S. M., Cohen, P., 1995. Inactivation of p42 MAP kinase by protein phosphatase 2A and a protein tyrosine phosphatase, but not CL100, in various cell lines. Curr. Biol. 5, 283–295.

Alon, U., Surette, M. G., Barkai, N., Leibler, S., 1999. Robustness in bacterial chemotaxis. Nature 397, 168–171.

Alonso, A., Sasin, J., Bottini, N., Friedberg, I., Friedberg, I., Osterman, A., Godzik, A., Hunter, T., Dixon, J., Mustelin, T., 2004. Protein tyrosine phosphatases in the human genome. Cell 117, 699–711.

Angeli, D., Ferrell, J. E., Sontag, E. D., 2004. Detection of multistability, bifurcations, and hysteresis in a large class of biological positive-feedback systems. Proc. Natl. Acad. Sci. USA 101, 1822–1827.

Aon, M. A., Gomez-Casati, D. F., Iglesias, A. A., Cortassa, S., 2001. Ultrasensitivity in (supra)molecularly organized and crowded environments. Cell. Biol. Int. 25, 1091–1099.

Arkin, A., Ross, J., McAdams, H. H., 1998. Stochastic kinetic analysis of developmental pathway bifurcation in phage lambda-infected Escherichia coli cells. Genetics 149, 1633–1648.

Ashburner, M., Ball, C. A., Blake, J. A., Botstein, D., Butler, H., Cherry, J. M., Davis, A. P., Dolinski, K., Dwight, S. S., Eppig, J. T., Harris, M. A., Hill, D. P., Issel-Tarver, L., Kasarskis, A., Lewis, S., Matese, J. C., Richardson, J. E., Ringwald, M., Rubin, G. M., Sherlock, G., 2000. Gene Ontology: tool for the unification of biology. The Gene Ontology Consortium. Nat. Genet. 25, 25–29.

Asthagiri, A. R., Horwitz, A. F., Lauffenburger, D. A., 1999a. A rapid and sensitive quantitative kinase activity assay using a convenient 96-well format. Anal. Biochem. 269, 342–347.

Asthagiri, A. R., Lauffenburger, D. A., 2001. A computational study of feedback effects on signal dynamics in a mitogen-activated protein kinase (MAPK) pathway model. Biotechnol. Prog. 17, 227–239.

Asthagiri, A. R., Nelson, C. M., Horwitz, A. F., Lauffenburger, D. A., 1999b. Quantitative relationship among integrin-ligand binding, adhesion, and signaling via focal adhesion kinase and extracellular signal-regulated kinase 2. J. Biol. Chem. 274, 27119–27127.

Asthagiri, A. R., Reinhart, C. A., Horwitz, A. F., Lauffenburger, D. A., 2000. The role of transient ERK2 signals in fibronectin- and insulin-mediated DNA synthesis. J. Cell. Sci. 113 Pt 24, 4499–4510.

Bagowski, C. P., Besser, J., Frey, C. R., Ferrell, J. E., 2003. The JNK cascade as a biochemical switch in mammalian cells: ultrasensitive and all-or-none responses. Curr. Biol. 13, 315–320.

Bagowski, C. P., Ferrell, J. E., 2001. Bistability in the JNK cascade. Curr. Biol. 11, 1176–1182.

Bandapalli, O. R., Geheeb, M., Kobelt, D., Kühnle, K., Elezkurta, S., Herrmann, J., Gressner, A. M., Weiskirchen, R., Beule, D., Blüthgen, N., Herzel, H., Franke, C., Brand, K., 2005. Global analysis of host tissue gene expression in the invasive front of colorectal liver metastases. International journal of cancer 118, 74–89.

Bard, J. B., Rhee, S. Y., 2004. Ontologies in biology: design, applications and future challenges. Nat. Rev. Genet. 5, 213–222.

Barkai, N., Leibler, S., 1997. Robustness in simple biochemical networks. Nature 387, 913–917.

Becskei, A., Seraphin, B., Serrano, L., 2001. Positive feedback in eukaryotic gene networks: cell differentiation by graded to binary response conversion. EMBO J. 20, 2528–2535.

Becskei, A., Serrano, L., 2000. Engineering stability in gene networks by autoregulation. Nature 405, 590–593.

Beinke, S., Robinson, M. J., Hugunin, M., Ley, S. C., 2004. Lipopolysaccharide activation of the TPL-2/MEK/extracellular signal-regulated kinase mitogen-activated protein kinase cascade is regulated by IkappaB kinase-induced proteolysis of NF-kappaB1 p105. Mol. Cell. Biol. 24, 9658–9667.

Beissbarth, T., Speed, T. P., 2004. GOstat: find statistically overrepresented Gene Ontologies within a group of genes. Bioinformatics 20, 1464–1465.

Benjamini, Y., Hochberg, Y., 1995. Controlling the false discovery rate: a practical and powerful approach to multiple testing. J. Roy. Statist. Soc. B 57, 289–300.

Benjamini, Y., Yekutieli, D., 2001. The control of the false discovery rate in multiple testing under dependency. Ann. Stat. 29, 1165–1188.

Bhalla, U. S., 2004. Signaling in small subcellular volumes. II. Stochastic and diffusion effects on synaptic network properties. Biophys. J. 87, 745–753.

Bhalla, U. S., Iyengar, R., 1999. Emergent properties of networks of biological signaling pathways. Science 283, 381–387.

Bhalla, U. S., Ram, P. T., Iyengar, R., 2002. MAP kinase phosphatase as a locus of flexibility in a mitogen-activated protein kinase signaling network. Science 297, 1018–1023.

Biggar, S. R., Crabtree, G. R., 2001. Cell signaling can direct either binary or graded transcriptional responses. EMBO J. 20, 3167–3176.

Binder, B., Heinrich, R., 2004. Interrelations between dynamical properties and structural characteristics of signal transduction networks. Genome Informatics 15, 13–23.

Birney, E., Andrews, D., Bevan, P., Caccamo, M., Cameron, G., Chen, Y., Clarke, L., Coates, G., Cox, T., Cuff, J., Curwen, V., Cutts, T., Down, T., Durbin, R., Eyras, E., Fernandez-Suarez, X. M., Gane, P., Gibbins, B., Gilbert, J., Hammond, M., Hotz, H., Iyer, V., Kahari, A., Jekosch, K., Kasprzyk, A., Keefe, D., Keenan, S., Lehvaslaiho, H., McVicker, G., Melsopp, C., Meidl, P., Mongin, E., Pettett, R., Potter, S., Proctor, G., Rae, M., Searle, S., Slater, G., Smedley, D., Smith, J., Spooner, W., Stabenau, A., Stalker, J., Storey, R., Ureta-Vidal, A., Woodwark, C., Clamp, M., Hubbard, T., 2004. Ensembl 2004. Nucleic Acids Res. 32 Database issue, D468–D470.

Blüthgen, N., 2002. Dynamical models of signal transduction and the influence of feedback loops. Diploma thesis, TU Berlin.

Blüthgen, N., Brand, K., Cajavec, B., Swat, M., Herzel, H., Beule, D., 2005a. Biological profiling utilizing gene ontology. Genome Informatics 16, 106–115.

Blüthgen, N., Bruggeman, F. J., Legewie, S., Herzel, H., Westerhoff, H. V., Kholodenko, B. N., 2005b. Effects of sequestration on signal transduction cascades. FEBS J. 273, 895–906.

Blüthgen, N., Herzel, H., 2001. MAP-kinase-cascade: Switch, amplifier or feedback controller. In: Gauges, R., van Gend, C., Kummer, U. (eds.), 2nd Workshop on Computation of Biochemical Pathways and Genetic Networks. pp. 55–62.

Blüthgen, N., Herzel, H., 2003. How robust are switches in intracellular signaling cascades? J. Theor. Biol. 225, 293–300.

Blüthgen, N., Kielbasa, S. M., Beule, D., 2006a. Handling, comparing and interpreting gene groups. In: Choi, S. (ed.), Introduction to Systems Biology, Humana Press, in press.

Blüthgen, N., Kielbasa, S. M., Cajavec, B., Herzel, H., 2004. HOMGL-comparing genelists across species and with different accession numbers. Bioinformatics 20, 125–126.

Blüthgen, N., Kielbasa, S. M., Herzel, H., 2005c. Inferring combinatorial regulation of transcription in silico. Nucleic Acids Res. 33, 272–279.

Blüthgen, N., Legewie, S., Herzel, H., Kholodenko, B., 2006b. Mechanisms generating ultrasensitivity, bistability and oscillation in signal transduction. In: Choi, S. (ed.), Introduction to Systems Biology, Humana Press, in press.

Bohr, C., Hasselbach, K. A., Krough, A., 1904. über einen in biologischen beziehungen wichtigen einfluß, den die kohlensäurespannung des blutes auf dessen sauerstoffbindung übt. Skand. Arch. Physiol. 16, 402–412.

Bos, J. L., 1989. Ras oncogenes in human cancer: a review. Cancer Res. 49, 4682–4689.

Bradshaw, J. M., Kubota, Y., Meyer, T., Schulman, H., 2003. An ultrasensitive Ca2+/calmodulin-dependent protein kinase II-protein phosphatase 1 switch facilitates specificity in postsynaptic calcium signaling. Proc. Natl. Acad. Sci. USA 100, 10512–10517.

Brightman, F. A., Fell, D. A., 2000. Differential feedback regulation of the MAPK cascade underlies the quantitative differences in EGF and NGF signalling in PC12 cells. FEBS Lett. 482, 169–174.

Brondello, J. M., Brunet, A., Pouyssegur, J., McKenzie, F. R., 1997. The dual specificity mitogen-activated protein kinase phosphatase-1 and -2 are induced by the p42/p44MAPK cascade. J. Biol. Chem. 272, 1368–1376.

Brown, G. C., Hoek, J. B., Kholodenko, B. N., 1997. Why do protein kinase cascades have more than one level? Trends. Biochem. Sci. 22, 288.

Brown, G. C., Kholodenko, B. N., 1999. Spatial gradients of cellular phosphoproteins. FEBS Lett. 457, 452–454.

Burd, A. L., Ingraham, R. H., Goldrick, S. E., Kroe, R. R., Crute, J. J., Grygon, C. A., 2004. Assembly of major histocompatability complex (MHC) class II transcription factors: association and promoter recognition of RFX proteins. Biochemistry 43, 12750–12760.

Burnham, K. P., Anderson, D. R., 1998. Model selection and inference - a practical information-theoretical approach. Springer, New York.

Campbell, S. L., Khosravi-Far, R., Rossman, K. L., Clark, G. J., Der, C. J., 1998. Increasing complexity of Ras signaling. Oncogene 17, 1395–1413.

Cardenas, M. L., 1997. Kinetic behaviour of vertebrate hexokinases with emphasis on hexokinase D (IV). Biochem. Soc. Trans. 25, 131–135.

Cardenas, M. L., Cornish-Bowden, A., 1989. Characteristics necessary for an interconvertible enzyme cascade to generate a highly sensitive response to an effector. Biochem. J. 257, 339–345.

Castillo-Davis, C. I., Hartl, D. L., 2003. GeneMerge–post-genomic analysis, data mining, and hypothesis testing. Bioinformatics 19, 891–892.

Cawley, S., Bekiranov, S., Ng, H. H., Kapranov, P., Sekinger, E. A., Kampa, D., Piccolboni, A., Sementchenko, V., Cheng, J., Williams, A. J., Wheeler, R., Wong, B., Drenkow, J., Yamanaka, M., Patel, S., Brubaker, S., Tammana, H., Helt, G., Struhl, K., Gingeras, T. R., 2004. Unbiased mapping of transcription factor binding sites along human chromosomes 21 and 22 points to widespread regulation of noncoding RNAs. Cell 116, 499–509.

Cho, K. H., Shin, S. Y., Kim, H. K., Wolkenhauer, O., McFerran, B., Kolch, W., 2003. Mathematical Modelling of the Influence of RKIP on the ERK Signalling Pathway. In: C.Priami (ed.), Computational Methods in Systems Biology. Springer, volume 2602 of Lecture Notes in Computer Science (LNCS), pp. 127–141.

Cohen, P., 2000. The regulation of protein function by multisite phosphorylation – a 25 year update. Trends Biochem. Sci. 25, 596–601.

Cohen, P., 2001. The role of protein phosphorylation in human health and disease. The Sir Hans Krebs Medal Lecture. Eur. J. Biochem. 268, 5001–5010.

Colman-Lerner, A., Gordon, A., Serra, E., Chin, T., Resnekov, O., Endy, D., Gustavo Pesce, C., Brent, R., 2005. Regulated cell-to-cell variation in a cell-fate decision system. Nature 437, 699–706.

Cuschieri, J., Bulmus, V., Gourlay, D., Garcia, I., Hoffman, A., Stayton, P., Maier, R. V., 2004. Modulation of macrophage responsiveness to lipopolysaccharide by IRAK-1 manipulation. Shock 21, 182–188.

Cyert, M. S., 2001. Regulation of nuclear localization during signaling. J. Biol. Chem. 276, 20805–20808.

Davis, R. J., 1995. Transcriptional regulation by MAP kinases. Mol. Reprod. Dev. 42, 459–467.

Davis, S., Vanhoutte, P., Pages, C., Caboche, J., Laroche, S., 2000. The MAPK/ERK cascade targets both Elk-1 and cAMP response element-binding protein to control long-term potentiation-dependent gene expression in the dentate gyrus in vivo. J. Neurosci. 20, 4563–4572.

De Cesare, D., Jacquot, S., Hanauer, A., Sassone-Corsi, P., 1998. Rsk-2 activity is necessary for epidermal growth factor-induced phosphorylation of CREB protein and transcription of c-fos gene. Proc. Natl. Acad. Sci. USA 95, 12202–12207.

Dennis, G., Sherman, B. T., Hosack, D. A., Yang, J., Gao, W., Lane, H. C., Lempicki, R. A., 2003. DAVID: Database for Annotation, Visualization, and Integrated Discovery. Genome Biol. 4, P3.

Deshaies, R. J., Ferrell, J. E., 2001. Multisite phosphorylation and the countdown to S phase. Cell 107, 819–822.

Dieterich, C., Cusack, B., Wang, H., Rateitschak, K., Krause, A., Vingron, M., 2002. Annotating regulatory DNA based on man-mouse genomic comparison. Bioinformatics 18 Suppl 2, S84–S90.

Dieterich, C., Wang, H., Rateitschak, K., Luz, H., Vingron, M., 2003. CORG: a database for comparative regulatory genomics. Nucleic Acids Res. 31, 55–57.

Dondi, E., Pattyn, E., Lutfalla, G., Ostade, X. v., Uze, G., Pellegrini, S., Tavernier, J., 2001. Down-modulation of type 1 interferon responses by receptor cross-competition for a shared Jak kinase. J. Biol. Chem. 276, 470004–47012.

Dorfman, K., Carrasco, D., Gruda, M., Ryan, C., Lira, S. A., Bravo, R., 1996. Disruption of the erp/mkp-1 gene does not affect mouse development: normal MAP kinase activity in ERP/MKP-1-deficient fibroblasts. Oncogene 13, 925–931.

Draghici, S., Khatri, P., Martins, R. P., Ostermeier, G. C., Krawetz, S. A., 2003. Global functional profiling of gene expression. Genomics 81, 98–104.

Dudoit, S., Shaffer, J. P., Boldrick, J. C., 2003. Multiple hypothesis testing in microarray experiments. Statistical Science 18, 71–103.

Dudoit, S., Yang, Y. H., Callow, M. J., Speed, T. P., 2002. Statistical methods for identifying differentially expressed genes in replicated cDNA microarray experiments. Statistica Sinica 12, 111–139.

Duggan, D. J., Bittner, M., Chen, Y., Meltzer, P., Trent, J. M., 1999. Expression profiling using cDNA microarrays. Nat. Genet. 21, 10–14.

Echevarria, D., Martinez, S., Marques, S., Lucas-Teixeira, V., Belo, J. A., 2005. Mkp3 is a negative feedback modulator of Fgf8 signaling in the mammalian isthmic organizer. Dev. Biol. 277, 114–128.

English, J. D., Sweatt, J. D., 1996. Activation of p42 mitogen-activated protein kinase in hippocampal long term potentiation. J. Biol. Chem. 271, 24329–24332.

Ermentrout, B., 2002. Simulating, Analyzing, and Animating Dynamical Systems: A Guide to XPPAUT for Researchers and Students. PA Society of Industrial and Applied Mathematics.

Euskirchen, G., Royce, T. E., Bertone, P., Martone, R., Rinn, J. L., Nelson, F. K., Sayward, F., Luscombe, N. M., Miller, P., Gerstein, M., Weissman, S., Snyder, M., 2004. CREB binds to multiple loci on human chromosome 22. Mol. Cell. Biol. 24, 3804–3814.

Farooq, A., Zhou, M. M., 2004. Structure and regulation of MAPK phosphatases. Cell. Signal. 16, 769–779.

Faust, K., 2005. Effects of Oncogenic Ras on Gene Expression: Clustering of Microarray Data and Screening for Potential Serum Response Factor Targets. Diploma thesis, Humboldt-Universität zu Berlin.

Fell, D. A., 1992. Metabolic control analysis: a survey of its theoretical and experimental development. Biochem. J. 286 (Pt 2), 313–330.

Fell, D. A., Sauro, H. M., 1990. Metabolic control analysis. The effects of high enzyme concentrations. Eur. J. Biochem. 192, 183–187.

Ferrell, J. E., 1996. Tripping the switch fantastic: how a protein kinase cascade can convert graded inputs into switch-like outputs. Trends Biochem. Sci. 21, 460–466.

Ferrell, J. E., 1998. How regulated protein translocation can produce switch-like responses. Trends. Biochem. Sci. 23, 461–465.

Ferrell, J. E., 2002. Self-perpetuating states in signal transduction: positive feedback, double-negative feedback and bistability. Curr. Opin. Cell. Biol. 14, 140–148.

Ferrell, J. E., Bhatt, R. R., 1997. Mechanistic studies of the dual phosphorylation of mitogen-activated protein kinase. J. Biol. Chem. 272, 19008–19016.

Ferrell, J. E., Machleder, E. M., 1998. The biochemical basis of an all-or-none cell fate switch in Xenopus oocytes. Science 280, 895–898.

Ferrell, J. E., Xiong, W., 2001. Bistability in cell signaling: How to make continuous processes discontinuous, and reversible processes irreversible. Chaos 11, 227–235.

Fessele, S., Boehlk, S., Mojaat, A., Miyamoto, N. G., Werner, T., Nelson, E. L., Schlondorff, D., Nelson, P. J., 2001. Molecular and in silico characterization of a promoter module and C/EBP element that mediate LPS-induced RANTES/CCL5 expression in monocytic cells. FASEB J. 15, 577–579.

Fessele, S., Maier, H., Zischek, C., Nelson, P. J., Werner, T., 2002. Regulatory context is a crucial part of gene function. Trends Genet. 18, 60–63.

Frith, M. C., Li, M. C., Weng, Z., 2003. Cluster-Buster: Finding dense clusters of motifs in DNA sequences. Nucleic Acids Res. 31, 3666–3668.

Furukawa, T., Sunamura, M., Motoi, F., Matsuno, S., Horii, A., 2003. Potential tumor suppressive pathway involving DUSP6/MKP-3 in pancreatic cancer. Am. J. Pathol. 162, 1807–1815.

Gardner, T. S., Cantor, C. R., Collins, J. J., 2000. Construction of a genetic toggle switch in Escherichia coli. Nature 403, 339–342.

Ge, Y., Jensen, T. L., Matherly, L. H., Taub, J. W., 2002. Synergistic regulation of human cystathionine-beta-synthase-1b promoter by transcription factors NF-YA isoforms and Sp1. Biochim. Biophys. Acta 1579, 73–80.

Gekakis, N., Staknis, D., Nguyen, H. B., Davis, F. C., Wilsbacher, L. D., King, D. P., Takahashi, J. S., Weitz, C. J., 1998. Role of the clock protein in the mammalian circadian mechanism. Science 280, 1564–1569.

Gentleman, R. C., Carey, V. J., Bates, D. M., Bolstad, B., Dettling, M., Dudoit, S., Ellis, B., Gautier, L., Ge, Y., Gentry, J., Hornik, K., Hothorn, T., Huber, W., Iacus, S., Irizarry, R., Li, F. L. C., Maechler, M., Rossini, A. J., Sawitzki, G., Smith, C., Smyth, G., Tierney, L., Yang, J. Y. H., Zhang, J., 2004. Bioconductor: Open software development for computational biology and bioinformatics. Genome Biology 5, R80.

Gibson, M. A., Bruck, J., 2000. Efficient exact stochastic simulation of chemical systems with many species and many channels. J. Phys. Chem. 104, 1876–1889.

Gillespie, D. T., 1977. Exact stochastic simulation of coupled chemical reactions. Journal of Physical Chem. 81, 2340–2361.

Goldberg, Y., 1999. Protein phosphatase 2A: who shall regulate the regulator? Biochem. Pharmacol. 57, 321–328.

Goldbeter, A., 1991. A minimal cascade model for the mitotic oscillator involving cyclin and cdc2 kinase. Proc. Natl. Acad. Sci. USA 88, 9107–9111.

Goldbeter, A., Koshland, D. E., 1981. An amplified sensitivity arising from covalent modification in biological systems. Proc. Natl. Acad. Sci. USA 78, 6840–6844.

Goldbeter, A., Koshland, D. E., 1984. Ultrasensitivity in biochemical systems controlled by covalent modification. Interplay between zero-order and multistep effects. J. Biol. Chem. 259, 14441–14447.

Gomez Casati, D. F., Aon, M. A., Iglesias, A. A., 1999. Ultrasensitive glycogen synthesis in Cyanobacteria. FEBS Lett. 446, 117–121.

Gong, Y., Zhao, X., 2003. Shc-dependent pathway is redundant but dominant in MAPK cascade activation by EGF receptors: a modeling inference. FEBS Lett. 554, 467–472.

Gu, Z., Kuntz-Simon, G., Rommelaere, J., Cornelis, J., 1999. Oncogenic transformation-dependent expression of a transcription factor NF-Y subunit. Mol. Carcinog. 24, 294–299.

Hartman, J. L., Garvik, B., Hartwell, L., 2001. Principles for the buffering of genetic variation. Science 291, 1001–1004.

Hasty, J., Pradines, J., Dolnik, M., Collins, J. J., 2000. Noise-based switches and amplifiers for gene expression. Proc. Natl. Acad. Sci. USA 97, 2075–2080.

Hatakeyama, M., Kimura, S., Naka, T., Kawasaki, T., Yumoto, N., Ichikawa, M., Kim, J. H., Saito, K., Saeki, M., Shirouzu, M., Yokoyama, S., Konagaya, A., 2003. A computational model on the modulation of mitogen-activated protein kinase (MAPK) and Akt pathways in heregulin-induced ErbB signalling. Biochem. J. 373, 451–463.

Heinemeyer, T., Wingender, E., Reuter, I., Hermjakob, H., Kel, A. E., Kel, O. V., Ignatieva, E. V., Ananko, E. A., Podkolodnaya, O. A., Kolpakov, F. A., Podkolodny, N. L., Kolchanov, N. A., 1998. Databases on transcriptional regulation: TRANSFAC, TRRD and COMPEL. Nucleic Acids Res. 26, 362–367.

Heinrich, R., Neel, B. G., Rapoport, T. A., 2002. Mathematical models of protein kinase signal transduction. Mol. Cell 9, 957–970.

Heinrich, R., Rapoport, T. A., 1974. A linear steady-state treatment of enzymatic chains. critique of the crossover theorem and a general procedure to identify interaction sites with an effector. Eur. J. Biochem. 42, 97–105.

Hill, A. V., 1910. The possible effects of the aggregation of the molecules of haemoglobin on its dissociation curves. J. Physiol. 40, iv–vii.

Hofer, T., Nathansen, H., Lohning, M., Radbruch, A., Heinrich, R., 2002. GATA-3 transcriptional imprinting in Th2 lymphocytes: a mathematical model. Proc. Natl. Acad. Sci. USA 99, 9364–9368.

Holmberg, C. I., Tran, S. E., Eriksson, J. E., Sistonen, L., 2002. Multisite phosphorylation provides sophisticated regulation of transcription factors. Trends Biochem. Sci. 27, 619–627.

Hornberg, J. J., Binder, B., Bruggeman, F. J., Schoeberl, B., Heinrich, R., Westerhoff, H. V., 2005a. Control of MAPK signalling: from complexity to what really matters:. Oncogene 24, 5533–5542.

Hornberg, J. J., Bruggeman, F. J., Binder, B., Geest, C. R., de Vaate, A. J., Lankelma, J., Heinrich, R., Westerhoff, H. V., 2005b. Principles behind the multifarious control of signal transduction. ERK phosphorylation and kinase/phosphatase control. FEBS J. 272, 244–258.

Hornberg, J. J., Tijssen, M. R., Lankelma, J., 2004. Synergistic activation of signalling to extracellular signal-regulated kinases 1 and 2 by epidermal growth factor and 4 beta-phorbol 12-myristate 13-acetate. Eur. J. Biochem. 271, 3905–3913.

Hosack, D. A., Dennis, Jr, G., Sherman, B. T., Lane, H. C., Lempicki, R. A., 2003. Identifying biological themes within lists of genes with EASE. Genome Biol. 4, R70.

Huang, C. Y., Ferrell, J. E., 1996. Ultrasensitivity in the mitogen-activated protein kinase cascade. Proc. Natl. Acad. Sci. USA 93, 10078–10083.

Huber, W., von Heydebreck, A., Sultmann, H., Poustka, A., Vingron, M., 2002. Variance stabilization applied to microarray data calibration and to the quantification of differential expression. Bioinformatics 18 Suppl 1, S96–104.

Jürchott, K., Kuban, R. J., Krech, T., Friese, C., Blüthgen, N., Stein, U., Lund, P., Valtin, H., Royer, H. D., Walther, W., Herzel, H., Schäfer, R., 2005. A crucial role of y-box binding proteins in oncogenic ras-mediated cell cycle control. under preparation .

Kacser, H., Burns, J. A., 1973. The control of flux. Symp. Soc. Exp. Biol 210, 65–104.

Kahn, D., Westerhoff, H. V., 1991. Control theory of regulatory cascades. J. Theor. Biol. 153, 255–285.

Kawakami, Y., Rodriguez-Leon, J., Koth, C. M., Buscher, D., Itoh, T., Raya, A., Ng, J. K., Esteban, C. R., Takahashi, S., Henrique, D., Schwarz, M. F., Asahara, H., Izpisua Belmonte, J. C., 2003. MKP3 mediates the cellular response to FGF8 signalling in the vertebrate limb. Nat. Cell. Biol. 5, 513–519.

Kholodenko, B. N., 2000. Negative feedback and ultrasensitivity can bring about oscillations in the mitogen-activated protein kinase cascades. Eur. J. Biochem. 267, 1583–1588.

Kholodenko, B. N., Demin, O. V., Moehren, G., Hoek, J. B., 1999. Quantification of short term signaling by the epidermal growth factor receptor. J. Biol. Chem. 274, 30169–30181.

Kholodenko, B. N., Hoek, J. B., Brown, G. C., Westerhoff, H. V., 1998. Control analysis of cellular signal transduction pathways. In: Larsson, C., Pahlman, I. L., Gustafsson, L. (eds.), Biothermokinetics in the postgenomic era. Chalmers, Goeteborg, pp. 102–107.

Kholodenko, B. N., Hoek, J. B., Westerhoff, H. V., 2000. Why cytoplasmic signalling proteins should be recruited to cell membranes. Trends. Cell. Biol. 10, 173–178.

Kholodenko, B. N., Hoek, J. B., Westerhoff, H. V., Brown, G. C., 1997. Quantification of information transfer via cellular signal transduction pathways. FEBS Letters 414, 430–434.

Kielbasa, S. M., Blüthgen, N., Herzel, H., 2004a. Genome-wide analysis of functions regulated by sets of transcription factors. In: Giegerich, R., J., S. (eds.), German Conference on Bioinformatics 2004. Gesellschaft für Informatik, Bonn, volume P-53, pp. 105–113.

Kielbasa, S. M., Blüthgen, N., Sers, C., Schäfer, R., Herzel, H., 2004b. Prediction of cis-regulatory elements of coregulated genes. Genome Informatics 15, 117–124.

Klose, J., Nock, C., Herrmann, M., Stuhler, K., Marcus, K., Bluggel, M., Krause, E., Schalkwyk, L. C., Rastan, S., Brown, S. D., Bussow, K., Himmelbauer, H., Lehrach, H., 2002. Genetic analysis of the mouse brain proteome. Nat. Genet. 30, 385–393.

Koshland, D. E., Goldbeter, A., Stock, J. B., 1982. Amplification and adaptation in regulatory and sensory systems. Science 217, 220–225.

Kufe, D. W., Pollock, R. E., Weichselbaum, R. R., Bast, Robert C., J., Gansler, T. S., Holland, J. F., Frei III, E. (Eds.), 2003. Cancer Medicine. BC Decker, Hamilton (Canada), 6th edition.

Langlois, W. J., Sasaoka, T., Saltiel, A. R., Olefsky, J. M., 1995. Negative feedback regulation and desensitization of insulin- and epidermal growth factor-stimulated p21ras activation. J. Biol. Chem. 270, 25320–25323.

LaPorte, D. C., Koshland, D. E., 1983. Phosphorylation of isocitrate dehydrogenase as a demonstration of enhanced sensitivity in covalent regulation. Nature 305, 286–290.

Lauffenburger, D. A., 2000. Cell signaling pathways as control modules: complexity for simplicity? Proc. Natl. Acad. Sci. USA 97, 5031–5033.

Lauffenburger, D. A., Linderman, J. J., 1993. Receptors – models for binding, trafficking, and signaling. Oxford University Press.

Lee, S., Park, J., Park, E., Chung, C., Kang, S., Bang, O., 2005. Akt-induced promotion of cell-cycle progression at G2/M phase involves upregulation of NF-Y binding activity in PC12 cells. J Cell Physiol 205, 270–277.

Legewie, S., Blüthgen, N., Herzel, H., 2005a. Quantitative analysis of ultrasensitive responses. FEBS J. 272, 4071–4079.

Legewie, S., Blüthgen, N., Schäfer, R., Herzel, H., 2005b. Supersensitization - switch-like regulation of cellular signal transmission by transcriptional induction. PLOS Comp. Biol. 1, e54.

Liu, H. S., Scrable, H., Villaret, D. B., Lieberman, M. A., Stambrook, P. J., 1992. Control of Ha-ras-mediated mammalian cell transformation by Escherichia coli regulatory elements. Cancer Res. 52, 983–989.

MacKeigan, J., Murphy, L., Dimitri, C., Blenis, J., 2005. Graded mitogen-activated protein kinase activity precedes switch-like c-Fos induction in mammalian cells. Mol Cell Biol 25, 4676–82.

Mangan, S., Zaslaver, A., Alon, U., 2003. The coherent feedforward loop serves as a sign-sensitive delay element in transcription networks. J Mol Biol 334, 197–204.

Manning, G., Whyte, D. B., Martinez, R., Hunter, T., Sudarsanam, S., 2002. The protein kinase complement of the human genome. Science 298, 1912–1934.

Marchetti, S., Gimond, C., Chambard, J. C., Touboul, T., Roux, D., Pouyssegur, J., Pages, G., 2005. Extracellular signal-regulated kinases phosphorylate mitogen-activated protein kinase phosphatase 3/DUSP6 at serines 159 and 197, two sites critical for its proteasomal degradation. Mol. Cell. Biol. 25, 854–864.

Markevich, N. I., Hoek, J. B., Kholodenko, B. N., 2004. Signaling switches and bistability arising from multisite phosphorylation in protein kinase cascades. J. Cell. Biol. 164, 353–359.

Marshall, C. J., 1995. Specificity of receptor tyrosine kinase signaling: transient versus sustained extracellular signal-regulated kinase activation. Cell 80, 179–185.

Martone, R., Euskirchen, G., Bertone, P., Hartman, S., Royce, T. E., Luscombe, N. M., Rinn, J. L., Nelson, F. K., Miller, P., Gerstein, M., Weissman, S., Snyder, M., 2003. Distribution of NF-kappaB-binding sites across human chromosome 22. Proc. Natl. Acad. Sci. USA 100, 12247–12252.

Mayawala, K., Gelmi, C. A., Edwards, J. S., 2004. MAPK cascade possesses decoupled controllability of signal amplification and duration. Biophys. J. 87, L01–2.

Meinke, M. H., Bishop, J. S., Edstrom, R. D., 1986. Zero-order ultrasensitivity in the regulation of glycogen phosphorylase. Proc. Natl. Acad. Sci. USA 83, 2865–2868.

Moehren, G., Markevich, N., Demin, O., Kiyatkin, A., Goryanin, I., Hoek, J. B., Kholodenko, B. N., 2002. Temperature dependence of the epidermal growth factor receptor signaling network can be accounted for by a kinetic model. Biochemistry 41, 306–320.

Murphy, L. O., Smith, S., Chen, R. H., Fingar, D. C., Blenis, J., 2002. Molecular interpretation of ERK signal duration by immediate early gene products. Nat. Cell. Biol. 4, 556–564.

Nash, P., Tang, X., Orlicky, S., Chen, Q., Gertler, F. B., Mendenhall, M. D., Sicheri, F., Pawson, T., Tyers, M., 2001. Multisite phosphorylation of a CDK inhibitor sets a threshold for the onset of DNA replication. Nature 414, 514–521.

Nelson, D. E., Ihekwaba, A. E., Elliott, M., Johnson, J. R., Gibney, C. A., Foreman, B. E., Nelson, G., See, V., Horton, C. A., Spiller, D. G., Edwards, S. W., McDowell, H. P., Unitt, J. F., Sullivan, E., Grimley, R., Benson, N., Broomhead, D., Kell, D. B., White, M. R., 2004. Oscillations in {NF-kappaB} signaling control the dynamics of gene expression. Science 306, 704–708.

O'Neill, E., Kolch, W., 2004. Conferring specificity on the ubiquitous Raf/MEK signalling pathway. Br. J. Cancer 90, 283–288.

Ortega, F., Acerenza, L., Westerhoff, H. V., Mas, F., Cascante, M., 2002. Product dependence and bifunctionality compromise the ultrasensitivity of signal transduction cascades. Proc. Natl. Acad. Sci. USA 99, 1170–1175.

Park, C. S., Schneider, I. C., Haugh, J. M., 2003. Kinetic analysis of platelet-derived growth factor receptor/phosphoinositide 3-kinase/Akt signaling in fibroblasts. J. Biol. Chem. 278, 37064–37072.

Pedraza, J. M., van Oudenaarden, A., 2005. Noise propagation in gene networks. Science 307, 1965–1969.

Pouyssegur, J., Volmat, V., Lenormand, P., 2002. Fidelity and spatio-temporal control in MAP kinase (ERKs) signalling. Biochem. Pharmacol. 64, 755–763.

Prudhomme, W., Daley, G. Q., Zandstra, P., Lauffenburger, D. A., 2004. Multivariate proteomic analysis of murine embryonic stem cell self-renewal versus differentiation signaling. Proc. Natl. Acad. Sci. USA 101, 2900–2905.

Raghavan, A., Ogilvie, R. L., Reilly, C., Abelson, M. L., Raghavan, S., Vasdewani, J., Krathwohl, M., Bohjanen, P. R., 2002. Genome-wide analysis of mRNA decay in resting and activated primary human T lymphocytes. Nucleic Acids Res. 30, 5529–5538.

Raser, J., O'Shea, E., 2004. Control of stochasticity in eukaryotic gene expression. Science 304, 1811–1814.

Ren, B., Cam, H., Takahashi, Y., Volkert, T., Terragni, J., Young, R. A., Dynlacht, B. D., 2002. E2F integrates cell cycle progression with DNA repair, replication, and G(2)/M checkpoints. Genes Dev. 16, 245–256.

Resat, H., Ewald, J. A., Dixon, D. A., Wiley, H. S., 2003. An integrated model of epidermal growth factor receptor trafficking and signal transduction. Biophys. J. 85, 730–743.

Rivera, V. M., Miranti, C. K., Misra, R. P., Ginty, D. D., Chen, R. H., Blenis, J., Greenberg, M. E., 1993. A growth factor-induced kinase phosphorylates the serum response factor at a site that regulates its DNA-binding activity. Mol. Cell. Biol. 13, 6260–6273.

Rosenfeld, N., Young, J. W., Alon, U., Swain, P. S., Elowitz, M. B., 2005. Gene regulation at the single-cell level. Science 307, 1962–1965.

Roulet, E., Busso, S., Camargo, A. A., Simpson, A. J., Mermod, N., Bucher, P., 2002. High-throughput SELEX SAGE method for quantitative modeling of transcription-factor binding sites. Nat. Biotechnol. 20, 831–835.

Sachs, K., Perez, O., Pe'er, D., Lauffenburger, D. A., Nolan, G. P., 2005. Causal protein-signaling networks derived from multiparameter single-cell data. Science 308, 523–529.

Salazar, C., Höfer, T., 2003. Allosteric regulation of the transcription factor NFAT1 by multiple phosphorylation sites: a mathematical analysis. J. Mol. Biol. 327, 31–45.

Salazar, C., Hofer, T., 2006. Kinetic models of phosphorylation cycles: a systematic approach using the rapid-equilibrium approximation for protein-protein interactions. Biosystems 83, 195–206.

Sandelin, A., Alkema, W., Engstrom, P., Wasserman, W. W., Lenhard, B., 2004. JASPAR: an open-access database for eukaryotic transcription factor binding profiles. Nucleic Acids Res. 32 Database issue, D91–D94.

Sasagawa, S., Ozaki, Y., Fujita, K., Kuroda, S., 2005. Prediction and validation of the distinct dynamics of transient and sustained ERK activation. Nat. Cell. Biol. 7, 365–373.

Sauro, H. M., 1994. Moiety-conserved cycles and metabolic control analysis: problems in sequestration and metabolic channelling. BioSystems 33, 55–67.

Sauro, H. M., Kholodenko, B. N., 2004. Quantitative analysis of signaling networks. Prog. Biophys. Mol. Biol. 86, 5–43.

Schafer, R., Tchernitsa, O. I., Zuber, J., Sers, C., 2003. Dissection of signal-regulated transcriptional modules by signaling pathway interference in oncogene-transformed cells. Adv. Enzyme. Regul. 43, 379–391.

Schneider, T. D., Stephens, R. M., 1990. Sequence logos: a new way to display consensus sequences. Nucleic Acids Res. 18, 6097–6100.

Schoeberl, B., Eichler-Jonsson, C., Gilles, E. D., Muller, G., 2002. Computational modeling of the dynamics of the MAP kinase cascade activated by surface and internalized EGF receptors. Nat. Biotechnol. 20, 370–375.

Sers, C., Tchernitsa, O. I., Zuber, J., Diatchenko, L., Zhumabayeva, B., Desai, S., Htun, S., Hyder, K., Wiechen, K., Agoulnik, A., Scharff, K. M., Siebert, P. D., Schafer, R., 2002. Gene expression profiling in RAS oncogene-transformed cell lines and in solid tumors using subtractive suppression hybridization and cDNA arrays. Adv. Enzyme Regul. 42, 63–82.

Shibata, T., Fujimoto, K., 2005. Noisy signal amplification in ultrasensitive signal transduction. Proc. Natl. Acad. Sci. USA 102, 331–336.

Shvartsman, S. Y., Muratov, C. B., Lauffenburger, D. A., 2002. Modeling and computational analysis of EGF receptor-mediated cell communication in Drosophila oogenesis. Development 129, 2577–2589.

Small, R., Fell, D. A., 1990. Covalent modification and metabolic control analysis. Eur. J. Biochem. 191, 405–411.

Stelling, J., Gilles, E. D., 2001. Robustness vs. identifiability of regulatory modules? In: Proc. of the 2nd international conference on systems biology (ICSB). pp. 181–190.

Stormo, G. D., 1998. Information content and free energy in DNA–protein interactions. J. Theor. Biol. 195, 135–137.

Swameye, I., Muller, T. G., Timmer, J., Sandra, O., Klingmuller, U., 2003. Identification of nucleocytoplasmic cycling as a remote sensor in cellular signaling by databased modeling. Proc. Natl. Acad. Sci. USA 100, 1028–1033.

Tchernitsa, O. I., Sers, C., Zuber, J., Hinzmann, B., Grips, M., Schramme, A., Lund, P., Schwendel, A., Rosenthal, A., Schafer, R., 2004. Transcriptional basis of KRAS oncogene-mediated cellular transformation in ovarian epithelial cells. Oncogene 23, 4536–4555.

Teusink, B., Passarge, J., Reijenga, C. A., Esgalhado, E., van der Weijden, C. C., Schepper, M., Walsh, M. C., Bakker, B. M., van Dam, K., Westerhoff, H. V., Snoep, J. L., 2000. Can yeast glycolysis be understood in terms of in vitro kinetics of the constituent enzymes? Testing biochemistry. Eur. J. Biochem. 267, 5313–5329.

Thattai, M., van Oudenaarden, A., 2004. Stochastic gene expression in fluctuating environments. Genetics 167, 523–530.

Thomas, R., Gathoye, A. M., Lambert, L., 1976. A complex control circuit. Regulation of immunity in temperate bacteriophages. Eur. J. Biochem. 71, 211–227.

Timmer, J., Müller, T. G., Swameye, I., Sandra, O., Klingmüller, U., 2004. Modelling the nonlinear dynamics of cellular signal transduction. Int. J. Bif. Chaos 14, 2069–2079.

Tsang, M., Dawid, I. B., 2004. Promotion and attenuation of FGF signaling through the Ras-MAPK pathway. Sci. STKE 2004, pe17.

Tullai, J. W., Schaffer, M. E., Mullenbrock, S., Kasif, S., Cooper, G. M., 2004. Identification of transcription factor binding sites upstream of human genes regulated by the phosphatidylinositol 3-kinase and MEK/ERK signaling pathways. J. Biol. Chem. 279, 20167–20177.

Tyson, J. J., Chen, K. C., Novak, B., 2003. Sniffers, buzzers, toggles and blinkers: dynamics of regulatory and signaling pathways in the cell. Curr. Opin. Cell. Biol. 15, 221–231.

Vayttaden, S. J., Ajay, S. M., Bhalla, U. S., 2004. A spectrum of models of signaling pathways. Chembiochem 5, 1365–1374.

Warmka, J. K., Mauro, L. J., Wattenberg, E. V., 2004. Mitogen-activated protein kinase phosphatase-3 is a tumor promoter target in initiated cells that express oncogenic Ras. J. Biol. Chem. 279, 33085–33092.

Wasserman, W. W., Krivan, W., 2003. In silico identification of metazoan transcriptional regulatory regions. Naturwissenschaften 90, 156–166.

Wasserman, W. W., Sandelin, A., 2004. Applied bioinformatics for the identification of regulatory elements. Nat. Rev. Genet. 5, 276–287.

Wellbrock, C., Karasarides, M., Marais, R., 2004. The RAF proteins take centre stage. Nat. Rev. Mol. Cell. Biol. 5, 875–885.

Werner, T., Fessele, S., Maier, H., Nelson, P., 2003. Computer modeling of promoter organization as a tool to study transcriptional coregulation. FASEB J. 17, 1228–1237.

Whitehurst, A., Cobb, M. H., White, M. A., 2004. Stimulus-coupled spatial restriction of extracellular signal-regulated kinase 1/2 activity contributes to the specificity of signal-response pathways. Mol. Cell. Biol. 24, 10145–10150.

Whitfield, M. L., Sherlock, G., Saldanha, A. J., Murray, J. I., Ball, C. A., Alexander, K. E., Matese, J. C., Perou, C. M., Hurt, M. M., Brown, P. O., Botstein, D., 2002. Identification of genes periodically expressed in the human cell cycle and their expression in tumors. Mol. Biol. Cell. 13, 1977–2000.

Wolf, J., Heinrich, R., 2000. Effect of cellular interaction on glycolytic oscillations in yeast: a theoretical investigation. Biochem. J. 345, 321–334.

Woolf, P. J., Prudhomme, W., Daheron, L., Daley, G. Q., Lauffenburger, D. A., 2005. Bayesian analysis of signaling networks governing embryonic stem cell fate decisions. Bioinformatics 21, 741–753.

Xiong, W., Ferrell, J. E., 2003. A positive-feedback-based bistable 'memory module' that governs a cell fate decision. Nature 426, 460–465.

Xu, S., Khoo, S., Dang, A., Witt, S., Do, V., Zhen, E., Schaefer, E. M., Cobb, M. H., 1997. Differential regulation of mitogen-activated protein/ERK kinase (MEK)1 and MEK2 and activation by a Ras-independent mechanism. Mol. Endocrinol. 11, 1618–1625.

Yeung, K., Seitz, T., Li, S., Janosch, P., McFerran, B., Kaiser, C., Fee, F., Katsanakis, K. D., Rose, D. W., Mischak, H., Sedivy, J. M., Kolch, W., 1999. Suppression of Raf-1 kinase activity and MAP kinase signalling by RKIP. Nature 401, 173–177.

Yi, T. M., Huang, Y., Simon, M. I., Doyle, J., 2000. Robust perfect adaptation in bacterial chemotaxis through integral feedback control. Proc. Natl. Acad. Sci. USA 97, 4649–4653.

Zak, D. E., Gonye, G. E., Schwaber, J. S., Doyle, F. J., 2003. Importance of input perturbations and stochastic gene expression in the reverse engineering of genetic regulatory networks: insights from an identifiability analysis of an in silico network. Genome Res. 13, 2396–2405.

Zeeberg, B. R., Feng, W., Wang, G., Wang, M. D., Fojo, A. T., Sunshine, M., Narasimhan, S., Kane, D. W., Reinhold, W. C., Lababidi, S., Bussey, K. J., Riss, J., Barrett, J. C., Weinstein, J. N., 2003. GoMiner: a resource for biological interpretation of genomic and proteomic data. Genome Biol. 4, R28.

Zhao, Q., Shepherd, E., Manson, M., Nelin, L., Sorokin, A., Liu, Y., 2005. The role of mitogen-activated protein kinase phosphatase-1 in the response of alveolar macrophages to lipopolysaccharide: attenuation of proinflammatory cytokine biosynthesis via feedback control of p38. J Biol Chem 280, 8101–8108.

Zhao, Y., Zhang, Z. Y., 2001. The mechanism of dephosphorylation of extracellular signal-regulated kinase 2 by mitogen-activated protein kinase phosphatase 3. J. Biol. Chem. 276, 32382–32391.

Zhong, S., Li, C., Wong, W. H., 2003. ChipInfo: Software for extracting gene annotation and gene ontology information for microarray analysis. Nucleic Acids Res. 31, 3483–3486.

Zuber, J., Tchernitsa, O. I., Hinzmann, B., Schmitz, A. C., Grips, M., Hellriegel, M., Sers, C., Rosenthal, A., Schafer, R., 2000. A genome-wide survey of RAS transformation targets. Nat. Genet. 24, 144–152.

Acknowledgements

Zuallererst möchte ich meinem Betreuer, **Hanspeter Herzel,** für seine erstklassige Betreuung und für die Möglichkeit, in seiner Gruppe zu arbeiten danken. Hanspeter war immer da, wenn ich ihn brauchte, und hat mir bei allen inhaltlichen oder administrativen Problemen Lösungen aufgezeigt.

Mein Dank gilt auch besonders meinen experimentellen Kollegen im SFB, insbesondere **Christine Sers, Reinhold Schäfer, Anja Schramme, Jana Keil, Karoline Faust** und **Karsten Jürchott,** für die tolle Zusammenarbeit, durch die ich gelernt habe, viel biologischer zu denken.

I thank my co-PhD-students **Szymon Kielbasa** and **Stefan Legewie** for both their friendship and the interesting projects we did together. Without Stefans never-ending knowledge on the bio-medical literature that he shared with me, this thesis would look different. I also had a excellent time with my other colleagues in ITB, including **Stephan Beirer, Maciej Swat, Branka Cajavec, Didier Gonze, Jana Wolf** and **Samuel Bernard.** And thank you for the sausage, Szymon and Maciej! Ich danke auch den ITB System-Managern **Andreas Hantschmann** und **Christian Waltermann** dafür, dass unser System immer lief und Probleme in kürzester Zeit behoben waren. Ich weiss, was das für eine harte Arbeit ist! Und ich danke **Karin Winkelhöfer** für ihre Hilfe in allen administrativen Dingen.

Dieter Beule habe ich neben vielen guten Diskussionen und Projekten auch ein spannendes Seminar und einige lustige Bergwanderungen zu verdanken. Dieter muss ich ausserdem dafür danken, dass er mich überzeugt hat, im ITB Diplom zu machen. Dieter und **Johannes Schuchardt** schulde ich auch Dank dafür, dass sie mich durch interessante Bioinformatik-Projekte in Übergangszeiten über Wasser gehalten haben. Die Coding-Marathons werde ich nicht vergessen.

It was an excellent experience to work and discuss with **Frank Brugge-**

179

man, who has always tons of ideas and enthusiasm. I hope we will continue our collaboration in future, Frank!

I thank **Upinder Bhalla** for hosting me at NCBI. My visit there and the discussions with him deeply shaped my view on signal transduction. And besides this, my visit in India was an experience that I'll surely never forget. I will not forget how warm Upi and **Karan, Adithya, Ajay, Raghi, Jyoti,** and **Sachin** hosted me. And thank you, **Jyoti,** for the excellent trip to Hampi!

I had exciting discussions and collaborations with **Herbert Sauro** and **Boris Kholodenko.** I enjoyed the time I spend in Herbert's lab. Thank you for all the support, Herbert and Boris! I also thank **Mariko Hatakeyama** for offering me to host me at Riken, and all the support she gave me for the application.

I thank **Hans Westerhoff** for bringing up the question at the 1st MTBIO workshop why ultrasensitivity should be caused by multisite phosphorylation when zero-order ultrasensitivity provides a simple tool, which stimulated much of my work.

Jens Timmer möchte ich weiterhin für die Einladungen nach und Diskussionen in Freiburg danken. Meinen neuen Kollegen in der AG Kuhl möchte ich für die nette Aufnahme und die lustige Zeit seit Juli danken. **Reinhard Heinrich** danke ich für sein großes Engagement im Graduiertenkolleg, von dem ich durch zahllose Vorträge und Reisen, insbesondere aber durch die Förderung der Zusammenarbeit mit Frank, profitiert habe. **Andreas Herrmann** und **Thomas Korte** möchte ich für die Hilfe beim IF-Mikroskopieren danken. Außerdem danke ich **Thomas Maiwald,** der mich in mühevoller Arbeit in die Kunst der Modellselektion einführte.

Schließlich möchte ich meinen Eltern **Rita** und **Viktor,** außerdem **Nico** und **Simone** für ihre Unterstützung danken.